Buildings' Thermal Performance and Energy Efficiency for a Sustainable Construction

Buildings' Thermal Performance and Energy Efficiency for a Sustainable Construction

Guest Editors

Jingyuan Zhao

Xuan Ma

Qian Zhang

MDPI

Basel • Beijing • Wuhan • Barcelona • Belgrade • Novi Sad • Cluj • Manchester

Guest Editors
Jingyuan Zhao
School of Architecture
Chang'an University
Xi'an
China

Xuan Ma
School of Architecture
Chang'an University
Xi'an
China

Qian Zhang
Joint School of Design
and Innovation
Xi'an Jiaotong University
Xi'an
China

Editorial Office
MDPI AG
Grosspeteranlage 5
4052 Basel, Switzerland

This is a reprint of the Special Issue, published open access by the journal *Buildings* (ISSN 2075-5309), freely accessible at: https://www.mdpi.com/journal/buildings/special_issues/B86CB53847.

For citation purposes, cite each article independently as indicated on the article page online and as indicated below:

Lastname, A.A.; Lastname, B.B. Article Title. *Journal Name* **Year**, *Volume Number*, Page Range.

ISBN 978-3-7258-5431-8 (Hbk)
ISBN 978-3-7258-5432-5 (PDF)
https://doi.org/10.3390/books978-3-7258-5432-5

Contents

About the Editors

Jingyuan Zhao

Jingyuan Zhao is a Professor at the School of Architecture, Chang'an University, Xi'an, China. Her research interests include thermal building performance, green building technologies, and sustainable urban design. She has led and participated in several national and provincial research projects on low-carbon construction and environmental adaptation. She has published extensively in international journals and contributed to the development of technical standards in building energy efficiency.

Xuan Ma

Xuan Ma is an Associate Professor at the School of Architecture, Chang'an University, Xi'an, China. His research focuses on the urban built environment, thermal diagnostics, and low-carbon building design. He has coordinated interdisciplinary projects on human–environment interaction, urban climate resilience, and pollution–carbon synergy. He has published widely in SCI-indexed journals, holds multiple patents on building thermal performance diagnostics, and serves as a reviewer and editorial board member for international journals in the field of sustainable construction.

Qian Zhang

Qian Zhang is a Lecturer at the School of Humanities and Social Science, Xi'an Jiaotong University, Xi'an, China. His academic interests include environmental sociology, sustainable development, and the human dimensions of urban and architectural change. He has conducted research on public participation in urban design and the social implications of low-carbon policies. His work emphasizes integrating human-centered perspectives with technical solutions in the built environment.

buildings

MDPI

Editorial

Buildings' Thermal Performance and Energy Efficiency for Sustainable Construction

Jingyuan Zhao [1], Xuan Ma [1,*] and Qian Zhang [2]

[1] School of Architecture, Chang'an University, Xi'an 710061, China; zjyqtt@163.com
[2] School of Humanities and Social Science, Xi'an Jiaotong University, Xi'an 710049, China; qian.zhang@xjtu.edu.cn
* Correspondence: mxozil@yahoo.com

1. Context and Rationale

The accelerating challenges of climate change and the global pursuit of sustainable development have positioned building thermal performance and energy efficiency at the forefront of academic and practical discourse. As the building sector is responsible for a significant portion of global energy consumption and carbon emissions, improving thermal performance is not merely a technical issue but also a societal necessity. Against this backdrop, this Special Issue of *Buildings* brings together eight contributions addressing the intertwined goals of reducing energy demand, mitigating carbon emissions, and ensuring human comfort and resilience across diverse climates and building types. Collectively, these studies provide novel insights that range from microclimatic optimization and façade design strategies to large-scale urban analyses, thereby offering a comprehensive view of sustainable construction practices.

2. Theme, Advances, and Article Highlights

This Special Issue covers a spectrum of scales—from urban outdoor environments to building-level designs—showcasing integrated approaches that blend passive strategies, renewable energy utilization, advanced technologies, and lifecycle thinking.

Zhang et al. Contribution 1 explore how participatory design and cognitive alignment between users and designers can foster inclusive, age-friendly rural environments. The paper highlights that enhancing social sustainability can be effectively integrated with low-carbon development principles, thereby broadening the scope of sustainable construction.

Ma et al. Contribution 2 examine how elevated semi-open spaces enhance airflow and shading, improving pedestrian-level comfort in cold regions. The study integrates simulations and measurements, offering design guidance for public spaces in climates where extreme cold remains a challenge.

Liu et al. Contribution 3 address energy scarcity in cold mountain environments. Through a combination of passive solar design strategies and spatial optimization methods, the study demonstrates how settlement planning can reduce energy demands while enhancing local adaptability to harsh climates.

Khan and Ghiai Contribution 4 synthesize recent advances in façade innovation. The review highlights how material choice, surface texture, and reflective properties can significantly influence both indoor thermal performance and outdoor microclimates, contributing to holistic environmental comfort.

Zhang et al. Contribution 5 evaluate the mitigation potential of low-carbon technologies in an urban context. By modeling technology deployment scenarios in Xi'an, the

 https://doi.org/10.3390/buildings15183410

authors provide quantitative evidence of how innovations in building systems can reduce carbon emissions and accelerate progress toward regional climate goals.

Song et al. Contribution 6 introduce a morphological indicator to evaluate sunlight–shade dynamics. Their findings reveal a strong correlation between insolation patterns and pedestrian comfort, thereby providing urban designers with a practical metric for optimizing outdoor spaces.

Chen et al. Contribution 7 leverage CFD simulations to analyze wind flow in dense dormitory complexes. The study highlights how building arrangement and orientation can improve ventilation performance, which is crucial for both energy efficiency and health in collective living spaces.

Li et al. Contribution 8 investigate urbanization dynamics at a regional scale. By quantifying the coordination between land development, energy consumption, and carbon emissions, this work provides critical evidence for policy frameworks aimed at balancing economic growth with ecological sustainability.

Together, these eight contributions illustrate how advances at multiple scales—from material innovations and building design to urban planning and social participation—can be aligned to foster sustainable and climate-resilient construction practices.

3. Outlook and Gratitude

This Special Issue provides a comprehensive overview of cutting-edge research that integrates thermal performance improvement with energy efficiency, spanning from microclimatic analyses of outdoor spaces to regional carbon-emission dynamics. The contributions collectively highlight that sustainability in the built environment requires a multi-scalar approach: technical optimization of buildings, integration of low-carbon technologies, consideration of user needs, and systemic urban governance.

Looking ahead, future research should emphasize cross-scale integration, particularly linking building-level thermal optimization with urban-scale carbon accounting and renewable energy deployment. Moreover, strengthening the interface between technological innovation and human-centered design will be essential to achieve climate resilience and sustainable development.

We extend our sincere gratitude to all authors for their high-quality contributions, to the reviewers for their thoughtful feedback, and to the editorial team at *Buildings* for their professional support. We hope that this Special Issue will serve as a valuable reference for researchers, practitioners, and policymakers dedicated to advancing sustainable construction worldwide.

Conflicts of Interest: The authors declare no conflict of interest.

List of Contributions:

1. Zhang, Z.; Zhan, Z.; Li, Y. User–Designer Cognitive Synergy: Enhancing Age-Friendly Rural Public Space Design. *Buildings* **2025**, *15*, 3078. https://doi.org/10.3390/buildings15173078.
2. Ma, X.; Luo, Q.; Yan, F.; Lei, Y.; Lu, Y.; Chen, H.; Yang, Y.; Feng, H.; Zhou, M.; Ding, H.; et al. Evaluating the Microclimatic Performance of Elevated Open Spaces for Outdoor Thermal Comfort in Cold Climate Zones. *Buildings* **2025**, *15*, 2777. https://doi.org/10.3390/buildings15152777.
3. Liu, B.; Song, W.; Liu, Y.; Wang, C.; Song, J. Optimal Spatial Configuration for Energy and Solar Use in Alpine-Frigid Resettlement Communities. *Buildings* **2025**, *15*, 2691. https://doi.org/10.3390/buildings15152691.
4. Khan, Z.; Ghiai, M. Enhancing Outdoor Environmental Comfort: A Review of Façade-Surface Strategies and Microclimate Impacts. *Buildings* **2025**, *15*, 2829. https://doi.org/10.3390/buildings15162829.

5. Zhang, D.; Sun, L.; Zhang, Y.; Liu, T.; Gao, L.; Wang, F.; Qiao, X.; Liu, Y.; Zuo, J.; Wang, Y. The Effect of Low-Carbon Technology on Carbon Emissions Reduction in the Building Sector: A Case Study of Xi'an, China. *Buildings* **2025**, *15*, 1989. https://doi.org/10.3390/buildings15121989.
6. Song, J.; Liu, Y.; Chow, D.H.C.; Liu, B.; Cho, S. Correlation Between the Insolation Shadow Ratio and Thermal Comfort of Urban Outdoor Spaces in Residential Areas in Xi'an. *Buildings* **2025**, *15*, 1995. https://doi.org/10.3390/buildings15121995.
7. Chen, X.; Kang, H.; Zhao, J.; Liu, Q. Optimization Design Research of Architectural Layout and Morphology in Multi-Story Dormitory Areas Based on Wind Environment Analysis. *Buildings* **2025**, *15*, 1747. https://doi.org/10.3390/buildings15101747.
8. Li, Z.; Yu, Y.; Liu, B.; Zhang, X.; Li, T.; Shi, N.; Ren, Y. The Coupling Coordination Degree and Spatio-Temporal Divergence Between Land Urbanization and Energy Consumption Carbon Emissions of China's Yangtze River Delta Urban Agglomeration. *Buildings* **2025**, *15*, 1880. https://doi.org/10.3390/buildings15111880.

buildings

MDPI

Article

User–Designer Cognitive Synergy: Enhancing Age-Friendly Rural Public Space Design

Zhihuan Zhang [1], Ziqi Zhan [2] and Yongchang Li [3,*]

[1] School of Art and Design, Changzhou University, Changzhou 213100, China; s23101351099@smail.cczu.edu.cn
[2] College of Economics and Management, Nanjing Forest University, Nanjing 210018, China; ywzq@njfu.edu.cn
[3] College of Art Design, Nanjing Forest University, Nanjing 210018, China
* Correspondence: nanlinlyc@njfu.edu.cn

Abstract: As rural populations age at an accelerating pace, the role of public spaces in enhancing the quality of life and promoting social engagement among older adults has become increasingly important. However, a significant cognitive gap persists between the actual needs of elderly users and the intentions of designers, often resulting in suboptimal design outcomes and underutilized spaces. This study centers on the concept of user–designer cognitive synergy, aiming to establish a systematic framework to bridge this cognitive divide and improve the design quality of age-friendly rural public spaces. Employing Grounded Theory, the FKANO model, and the DEMATEL method, this study systematically elicited user needs, classified their attributes, and mapped causal relations to determine priority drivers. Applied in a representative rural case, the framework identified Environmental and Operations Management and Spatial Accessibility as the most critical needs, while Smart-Friendliness and Safety Organization were also shown to play significant roles. These findings directly informed targeted spatial strategies such as barrier-free circulation, robust nighttime safety systems, intergenerational hubs, and an operations backbone. Post-occupancy evaluation confirmed high satisfaction across safety, accessibility, functionality, social participation, and environmental comfort. The results demonstrate the framework's effectiveness in translating complex needs into actionable design strategies, offering both theoretical insights and practical guidance for age-friendly rural public space development.

Keywords: age-friendly design; rural public spaces; user–designer cognitive synergy; aging-in-place strategies

1. Introduction

With the rapid acceleration of global population aging, the design of age-friendly public spaces has become an urgent concern in both urban and rural contexts. In rural areas, public spaces play a particularly crucial role in supporting older adults' daily activities, social engagement, and cultural practices. However, limited design and management resources in these settings mean that even minor design oversights can have disproportionate impacts on accessibility, safety, and well-being.

Although existing studies have advanced knowledge of spatial layouts, accessibility, and environmental comfort in age-friendly design, they remain predominantly focused on physical functionality. For example, Han [1] emphasized spatial design as the key factor in elderly adaptability, while Feng [2] proposed community-based principles of mutual support in rural Shanghai. Ren et al. [3] demonstrated how specific spatial elements shape perceptual responses using eye-tracking technology. In international contexts, Nutt et al. [4] assessed Estonian towns and found that universal age-friendly principles were often unmet, and Hartt et al. [5] showed that older adult play spaces require structured yet flexible conditions to support diverse activities. In China, Gong et al. [6] identified outdoor space challenges in Hangzhou's residential neighborhoods, Zeng and Chen [7] confirmed safety and accessibility as priorities in Wuhan, Yi et al. [8] optimized healthcare facilities in Sanya to reduce walking distances and improve efficiency, and Cao et al. [9] highlighted the differentiated needs of elderly travelers in Nanjing railway station. These studies provide valuable methodological and scenario-specific insights, yet they mostly emphasize physical and environmental improvements. The diverse and layered needs of older adults, particularly those related to social interaction, emotional belonging, and cognitive alignment with designers, remain underexplored. Field investigations in Jiangning District, Nanjing, further illustrate these gaps: ramps exist only as isolated interventions without continuous accessibility, nighttime safety measures are insufficient, and benches without armrests are sited away from play areas, weakening intergenerational interaction. Such shortcomings highlight the absence of a systematic pathway from identifying user needs to translating them into prioritized interventions.

This study seeks to address these gaps by developing a User–Designer Cognitive Alignment Framework for rural age-friendly public spaces. The objectives are threefold: to uncover cognitive discrepancies between elderly users and designers in relation to spatial functions, environmental perception, and emotional attachment; to classify and prioritize user needs according to their impact on satisfaction; and to construct a causal mapping that translates cognitive findings into design strategies.

To achieve these objectives, the study integrates Grounded Theory (GT), the Fuzzy Kano (FKANO) model, and the Decision-Making Trial and Evaluation Laboratory (DEMATEL) method. Grounded Theory is employed to systematically elicit and code multidimensional user needs, the FKANO model is used to classify attributes and assess satisfaction elasticity, and DEMATEL is applied to reveal causal relationships and weightings among needs. This integrated approach ensures both qualitative depth and quantitative precision. Applied in representative rural cases in Jiangsu Province, the framework highlights critical design drivers such as barrier-free circulation, nighttime safety, and environmental operations management. The findings not only contribute to theoretical enrichment in participatory design research but also provide practical guidance for enhancing the livability and inclusivity of rural communities.

The remainder of this paper is structured as follows. Section 2 reviews related studies on age-friendly rural public spaces and user–designer cognition. Section 3 introduces the theoretical foundation and research methodology. Section 4 presents the empirical analysis, including need identification, classification, and causal mapping, followed by the proposed design intervention. Section 5 evaluates the feasibility of the intervention, and Section 6 concludes with key contributions and future research directions.

2. Literature Review

2.1. Age-Friendly Design in Public Spaces

With the intensification of global population aging, the development of age-friendly public spaces has become a key focus of academic research. In rural contexts, public spaces serve as essential venues for older adults' daily activities and social engagement, playing a vital role in supporting both their physical and mental well-being [10]. Current research primarily addresses age-friendly design through the lenses of universal accessibility, social participation, and environmental safety, with considerable effort devoted to developing evaluation frameworks for assessing the age-friendliness of public spaces. For instance, Helbich et al. [11] explored the relationship between older adults' exposure to green-blue spaces and depressive symptoms, finding that street-level green-blue spaces were negatively correlated with depression scores, while satellite-based indicators showed no significant association, highlighting the importance of perceived environmental quality. Lin et al. [12] developed an age-friendly seating evaluation system for urban parks based on the Kano–IPA model, identifying 20 indicators that influence older users' satisfaction. Zhang et al. [13] proposed a composite framework for evaluating the quantity, quality, and spatial distribution of community open spaces, introducing the concept of "adaptive spatial share" to assess intergenerational equity.

Additionally, Jiang et al. [14], through a survey of 108 elderly residents in Suzhou, demonstrated that community infrastructure, street network conditions, and environmental satisfaction positively influence physical activity and health outcomes, providing a theoretical basis for optimizing community facilities and street environments. Scholars have also proposed targeted strategies for different types of public spaces. Zhou et al. [15] proposed a three-dimensional framework for improving community interaction spaces—encompassing functional complexity, age-sensitivity, and precision in construction—such as enhancing seating comfort in conversational areas or integrating intergenerational interaction elements into fitness zones. Han et al. [16] developed a quantitative evaluation system for age-friendly design in affordable housing communities in Shenzhen using the Analytic Hierarchy Process (AHP). Gong et al. [17], analyzing the Gongchen Bridge waterfront section of the Grand Canal, used factor analysis to identify five key dimensions, with environmental factors having the strongest influence on elderly satisfaction. In cold-region rural settings, Han and Yu [18] applied a fuzzy AHP approach, combined with Delphi expert consultation, to construct an evaluation model for environmental carrying capacity in cold-climate rural public spaces, clarifying the roles of socioeconomic and spatial factors.

In summary, significant progress has been made in developing evaluation frameworks and urban-oriented strategies for age-friendly public space design. However, several critical gaps remain. First, most existing studies are urban-centric, overlooking the complexity and multidimensionality of rural public space use, where older adults rely more heavily on such spaces to fulfill social, ceremonial, and emotional needs. Second, the predominant reliance on quantitative methods limits insight into subjective experiences and perceptions; user needs are often inferred from observed behavior, neglecting subjective cognition and potential designer misinterpretations. Third, few studies explore systematic co-design mechanisms that bridge the cognitive divide between users and designers. To address these gaps, this study focuses specifically on age-friendly design in rural public spaces, adopting a multi-perspective, multi-stage methodology centered on user–designer collaboration, aiming to contribute both theoretical enrichment and practical pathways for context-sensitive and participatory design innovation.

2.2. Multi-Method Integration in Collaborative Design

In recent years, user participation and collaborative design have become central paradigms across design disciplines. In response to increasingly diverse and complex user demands, the field has shifted from expert-led models toward human-centered co-creation, emphasizing early user involvement to enhance design relevance, usability, and acceptance. For example, in the healthcare domain, Witteman et al. [19] developed the User-Centered Design 11-item Scale using psychometric methods, validating the positive impact of multi-stage user participation on tool usability. In age-friendly product design, Wang et al. [20] integrated the Kano model, AHP, and Fuzzy Comprehensive Evaluation to systematically address feature prioritization and solution optimization for elderly walkers. In service system design, Du [21] innovatively combined DEMATEL and TOPSIS to resolve stakeholder conflicts and manage risk compensation in segmented logistics services. For environments designed for special populations, Liu et al. [22] applied Axiomatic Design to translate the interactive needs of children with intellectual disabilities into spatial design parameters, demonstrating the value of layered demand modeling.

In the domain of intelligent infrastructure, Liu [23] developed an analytical framework by integrating the Kawakita Jiro Method, Kano model, and AHP to optimize core functions of electric vehicle charging stations. In the domain of home textile design, Qin et al. [24] explored the relationship between the symmetry of traditional Jiangnan homespun patterns and user perception. Emotional descriptors were gathered using affinity diagramming, and Principal Component Analysis (PCA) was employed to extract key perceptual dimensions. The Fuzzy Analytic Hierarchy Process (FAHP) was then used to assign relative weights to these dimensions. Based on the results, five pattern variants were reconstructed using shape grammar, and the optimal design scheme was identified through Quality Function Deployment (QFD) combined with Grey Relational Analysis (GRA). Likewise, Wang et al. [25] applied modularity and axiomatic design to achieve lifecycle-based adaptability in children's furniture design. Wang et al. [26] employed a combination of affinity diagramming, the Analytic Hierarchy Process (AHP), and the entropy weight method to investigate the relationship between emotional vocabulary, cultural symbols, and user perception. Drawing upon artifact elements from the tomb of the Marquis of Haihun, the authors extracted representative cultural features and, through Quality Function Deployment (QFD), established a structured mapping to inform the optimization of Chinese-style seating design. These studies share a common cross-disciplinary methodological structure, needs identification, attribute classification, quantitative prioritization, and design verification, offering a replicable model for collaborative design across various domains. Such frameworks have significantly expanded the theoretical foundations and methodological toolkits for participatory and demand-driven design.

Despite these advancements, several limitations persist in the context of age-friendly spatial design, particularly in rural settings. First, although multi-method integration has proven effective in product and service domains, its applicability to socially complex, spatially embedded rural public spaces remains underexplored. Second, many studies emphasize the user perspective, while overlooking the designer's cognitive framework, including their assumptions, knowledge structures, and interpretive filters. This creates a disconnect where "collaboration" becomes symbolic or superficial, lacking mechanisms for cognitive convergence and meaningful co-creation. In response, this study proposes a User–Designer Cognitive Alignment Framework that integrates Grounded Theory to reveal cognitive divergence, the FKANO model to classify and prioritize user needs, and the DEMATEL method to map causal relationships among needs. This integrated methodology enables structured modeling of collaborative design processes and provides empirical validation and actionable strategies for building consensus and supporting decision-making in complex age-friendly design contexts.

3. Methods

3.1. Research Framework and Process Design

This study aims to enhance the age-friendliness of rural public spaces by establishing a User–Designer Cognitive Alignment Framework, thereby addressing the multifaceted spatial needs of the elderly population. To achieve this goal, a comprehensive framework is proposed, integrating GT, the FKANO model, and the DEMATEL method, as illustrated in Figure 1. The framework proceeds through the following stages:

- Semi-structured interviews and field observations were conducted with elderly users and designers to capture cognitive perspectives regarding functions, environmental settings, and emotional associations. Through open, axial, and selective coding, authentic user needs were systematically extracted and organized into preliminary categories. These categories served as the direct item pool for the FKANO questionnaire design, ensuring that the subsequent quantitative analysis was firmly grounded in empirical data.

- Using the GT-derived needs as inputs, a structured questionnaire was distributed to classify each demand element into Must-be (M), One-dimensional (O), or Attractive (A) attributes, while also quantifying their influence on satisfaction through Better–Worse coefficients. This step revealed cognitive gaps between users and designers in interpreting and prioritizing needs. FKANO results provided the initial weighting cues and category typologies that guided the DEMATEL analysis.
- Expert evaluation was conducted on the FKANO-screened needs to build a Direct Influence Matrix. DEMATEL was then applied to uncover causal relationships, calculate prominence (D + C) and net influence (D − C), and identify key driving and dependent factors. Importantly, DEMATEL also acted as a validation step: Must-be and high-priority One-dimensional items identified in FKANO were expected to emerge as causal drivers with high prominence, whereas Attractive items were more likely to appear as dependent effects. Any inconsistencies were examined and reconciled through contextual interpretation, thereby cross-validating the robustness of the FKANO classification.

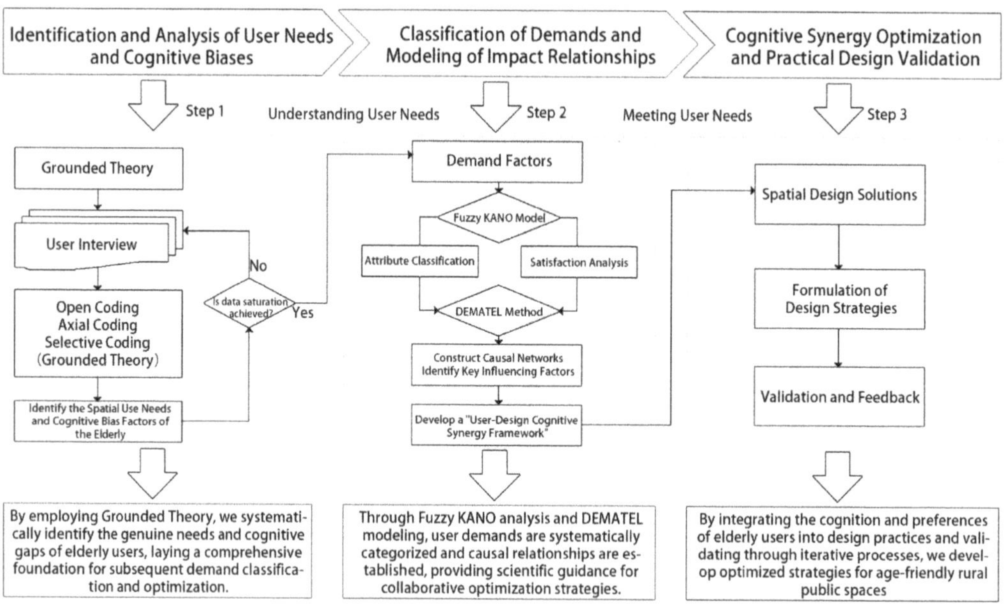

Figure 1. Proposed Framework.

By integrating these methods, the framework forms a closed-loop pathway from qualitative need elicitation (GT) to quantitative categorization and prioritization (FKANO) to causal validation and decision mapping (DEMATEL). This multi-method synergy ensures that design interventions are both empirically grounded and logically prioritized. A representative rural public space was then selected for design intervention, where iterative refinement translated theoretical findings into actionable spatial strategies, directly addressing neglected but high-impact user needs.

3.2. Grounded Theory

Grounded Theory (GT) is a qualitative methodology that emphasizes theory generation from empirical data rather than testing pre-existing hypotheses [27,28]. Originally proposed by Glaser and Strauss, GT has been widely applied in healthcare, sociology, and, more recently, in design research, where it serves as a rigorous tool for eliciting user needs and constructing theoretical frameworks [29]. Its advantage lies in uncovering latent patterns and building explanatory models that capture complex social realities [30].

In this study, GT was applied to identify the multidimensional needs of rural elderly users and to reveal cognitive mismatches between users and designers [31,32]. The process involved five steps:

(1) Data Collection: Semi-structured interviews and on-site observations were conducted with elderly users and spatial designers, focusing on functional requirements, safety, emotional and social dimensions, and environmental maintenance.
(2) Open Coding: Interview transcripts were analyzed line by line to extract key expressions, which were then labeled and grouped into initial conceptual categories.
(3) Axial Coding: Relationships among these concepts were examined to form broader themes such as functional needs, emotional belonging, and environmental management.
(4) Selective Coding: Core categories were consolidated into a coherent framework, highlighting areas of divergence or conflict between user expectations and design assumptions. These results also provided inputs for FKANO classification and DEMATEL analysis.
(5) Theoretical Saturation Testing: Additional interviews were used to confirm completeness; once no new categories emerged, the framework was deemed saturated and ready for subsequent model validation.

Through this process, GT ensured that the subsequent quantitative analyses were firmly grounded in authentic user cognition, thereby strengthening the reliability and contextual relevance of the research framework.

3.3. Fuzzy Kano Model

The Kano model is a qualitative tool widely used to categorize user needs and determine design priorities. It illustrates the nonlinear relationship between design features and user satisfaction, as shown in Figure 2 [33,34]. However, traditional Kano analysis cannot fully capture subjectivity and uncertainty in user feedback. The FKANO model addresses this by integrating fuzzy logic and fuzzy set theory, allowing for a more accurate evaluation of user expectations [35,36].

The FKANO model enables users to express satisfaction degrees using fractional values between 0 and 1 across different response options [37], better reflecting ambiguous real-world preferences [38]. In this study, FKANO is applied to classify and quantify age-friendly attributes in rural public space design. FKANO implementation involves four steps:

(1) Questionnaire Design and Distribution: A questionnaire based on GT-derived needs is created and distributed both online and offline to ensure coverage across diverse elderly user groups.
(2) Interaction Matrix Construction: Based on user responses, presence (P) and absence (N) matrices are constructed. From these, the interaction matrix S is derived, capturing how different attributes affect user attitudes:

$$S = P^T \times N \tag{1}$$

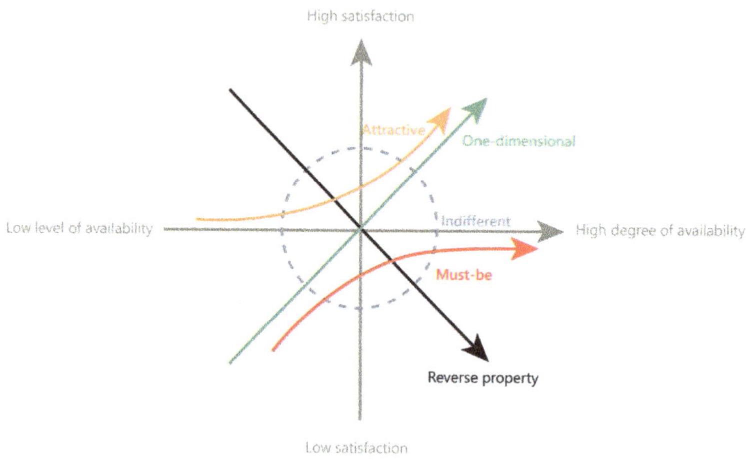

Figure 2. KANO Model.

(3) Attribute Classification via Membership Vector: The values in matrix *S* are mapped to an evaluation table (Table 1) to derive membership vectors T. Using a threshold of α = 0.4, attributes are classified as 1 or 0. If multiple categories are valid, they are prioritized as: Must-be (M) > One-dimensional (O) > Attractive (A) > Indifferent (I) > Reverse (R).

Table 1. FKANO Model Evaluation Matrix.

Perceived Requirement		If the Feature Is Absent				
		Likes	**Expects So**	**Neutral**	**Can Accept**	**Dislikes**
If this feature is present	Likes	*Q*	*A*	*A*	*A*	*O*
	Expects So	*R*	*I*	*I*	*I*	*M*
	Neutral	*R*	*I*	*I*	*I*	*M*
	Can Accept	*R*	*I*	*I*	*I*	*M*
	Dislikes	*R*	*R*	*R*	*R*	*Q*

Must-be requirements (*M*), One-dimensional requirements (O), Attractive requirements (*A*), Indifferent requirements (*I*), Reverse requirements (*R*).

(4) Satisfaction Impact Analysis: The proportions of A_i, M_i, O_i, and I_i are used to calculate two indicators:
Satisfaction Coefficient:

$$S_i = \frac{A_i + O_i}{A_i + O_i + M_i + I_i} \quad (2)$$

Dissatisfaction Coefficient:

$$D_i = -1 \cdot \frac{M_i + O_i}{A_i + O_i + M_i + I_i} \quad (3)$$

These coefficients support identification of high-priority features that significantly influence satisfaction and guide design trade-offs.

Linking sources to outputs and validation: (1) Item generation (GT → FKANO): Questionnaire items were directly derived from GT selective categories, formulated as paired positive/negative "presence/absence" statements. Each item is traceable to its originating open/axial code cluster, ensuring comprehensive need coverage. (2) Screening and weighting (FKANO → DEMATEL): Based on FKANO classes and Better–Worse coefficients, we prioritized Must-be (M) and high-sensitivity One-dimensional (O) items, and contextually admitted a subset of Attractive (A) items. FKANO outputs were translated into the DEMATEL factor list with initial importance references to guide expert pairwise weighting and threshold setting. (3) Cross-validation expectations: M and high-O items are expected to exhibit high prominence (D + C) and act as causal drivers (D − C > 0) in DEMATEL, whereas A items tend to be effect receivers (D − C < 0) with lower prominence. Any discrepancies are contextually reconciled and calibrated.

3.4. Decision-Making Trial and Evaluation Laboratory Method

As multi-attribute decision-making becomes increasingly complex, the DEMATEL method has gained popularity for visualizing and analyzing causal relationships among interrelated factors [39]. Its strength lies in mapping direct and indirect influences, enabling decision-makers to prioritize interventions with clarity and precision [40]. In this study, DEMATEL is applied to analyze interdependencies among user needs in rural age-friendly design [41], with the goal of identifying root causes [42], prioritizing interventions, and supporting evidence-based decisions [43]. The procedure includes:

(1) Pairwise Influence Evaluation: Experts rate the degree of influence each demand element has on others, generating a direct-relation matrix A.

$$A = \begin{bmatrix} 0 & a_{12} & \cdots & a_{1n} \\ a_{21} & 0 & \cdots & a_{2n} \\ \vdots & \vdots & 0 & \vdots \\ a_{n1} & a_{n2} & \cdots & 0 \end{bmatrix} = \left(A_{ij}\right)_{n\times n} \tag{4}$$

(2) Normalization of the Matrix: Matrix A is normalized to obtain matrix X, ensuring that the total influence score does not exceed 1.

$$X = A / \max_{1\le i\le n} \sum_{j=1}^{n} A_{ij} = \left(x_{ij}\right)_{n\times n} \tag{5}$$

(3) Total Influence Matrix Calculation: The total influence matrix T is computed using the following formula:

$$T = X(I - X)^{-1} \tag{6}$$

Capturing both direct and indirect effects.

(4) Causal Metrics Derivation: For each factor, calculate:
Degree of influence:

$$D_i = \sum_{j=1}^{n} T_{ij} (1 \le i \le n, 1 \le j \le n) \tag{7}$$

Degree of being influenced:

$$C_i = \sum_{j=1}^{n} T_{ij} (1 \le i \le n, 1 \le j \le n) \tag{8}$$

Analyze $D_i + C_i$ (prominence) and $D_i - C_i$ (net influence) to generate an Influence Network Relation Map (INRM) (Figure 3), categorizing demand elements.

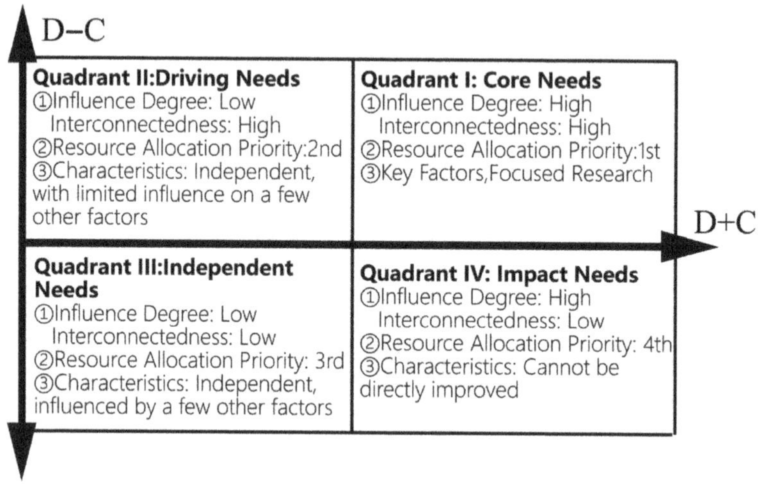

Figure 3. INRM-Based Classification of Demand Categories.

This method offers a structured approach to uncovering underlying causal chains, strengthening the systematicity, scientific validity, and practical alignment of rural age-friendly spatial design.

4. Case Study

4.1. User Need Identification Based on Grounded Theory

(1) Data Collection

A purposive sampling approach was adopted to recruit elderly residents from villages in Jiangning District, Nanjing, Jiangsu Province, for in-depth interviews conducted between March and May 2025. The final sample size was determined based on the principle of theoretical saturation, resulting in a total of 20 participants representing a diverse cross-section of the rural elderly population. Basic information about the interviewees is provided in Figure 4. Each participant participated in a face-to-face interview guided by a semi-structured protocol (Appendix A). The interviews, lasting approximately 20–30 min each, were fully audio-recorded. After each interview, transcripts were promptly compiled and returned to participants for review and verification, ensuring data accuracy and completeness.

(2) Open Coding

Following data collection, all interview transcripts were analyzed using the coding procedures of Grounded Theory, comprising open, axial, and selective coding stages. To enhance analytic reliability, three researchers in the field of design jointly participated in the coding process, ultimately producing a robust coding structure. During open coding, the textual data were reviewed line by line to extract key expressions related to user needs in rural public spaces. A total of 23 preliminary conceptual categories were identified from user narratives. These categories were then organized into an initial conceptual framework, forming the basis for more structured theoretical abstraction.

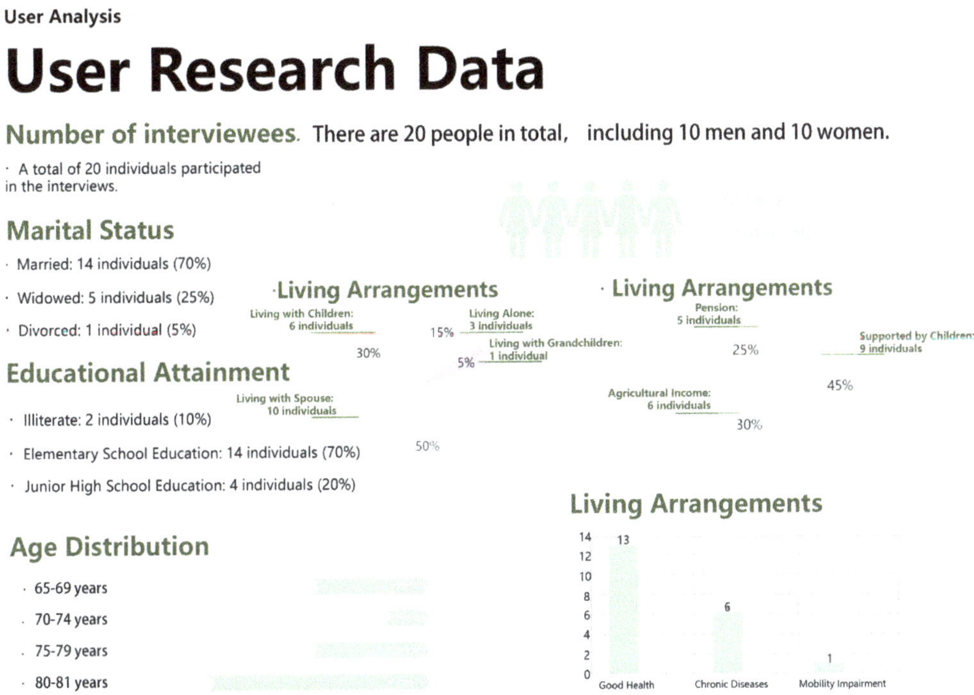

Figure 4. Background Information of Interview Participants.

(3) Axial Coding

In the axial coding phase, the 55 initial concepts identified during open coding were further classified and synthesized to explore their underlying logical relationships. Through systematic integration, the 23 conceptual categories were consolidated into four overarching axial categories, forming a more cohesive and hierarchical theoretical structure for analyzing age-friendly rural public space design. This process allowed for deeper insights into the internal linkages among user needs and clarified how these dimensions interact within the rural design context. The complete coding results are presented in Table 2, which lays a solid foundation for subsequent empirical analysis and theoretical modeling.

(4) Selective Coding

The selective coding phase focused on synthesizing the relationships among axial categories into a unified and systematic theoretical framework. Based on the results of the prior coding stages, the demand indicators were organized into key themes, functional attributes, emotional belonging, and environmental management, which reflect the experiential and perceptual needs of elderly users. The final outcome was a hierarchical model of elderly spatial demands, comprising four major categories: functional completeness, safety assurance, emotional attachment, and environmental governance, as depicted in Figure 5. This model provides a structured theoretical basis for the empirical validation and subsequent design optimization of rural age-friendly public spaces.

Table 2. Coding Results Based on Grounded Theory.

Core Category	Subcategory	Initial Concepts
Basic Functional and Smart Needs (C1)	Spatial Convenience (B1)	Proximity to users (A1), Convenient transportation (A2), High accessibility (A3)
	Facility Usability (B2)	Comfortable seating (A4), Appropriate fitness equipment (A5), Shelter from sun/rain (A6)
	Functional Diversity (B3)	Resting areas (A7), Recreational zones (A8), On-site medical facilities (A9), Drinking stations (A10)
	Smart-Friendliness (B4)	Facial recognition (A11), Mobile app operability (A12)
	Shared Internet Access (B5)	Full local network coverage (A13)
	Barrier-free Accessibility (B6)	No steps (A14), Wheelchair ramps (A15), Well-placed handrails (A16)
Safety and Security Needs (C2)	Facility Safety (B7)	Non-slip flooring (A17), Protective railings (A18)
	Emergency Response (B8)	Emergency medical equipment (A19), Emergency call buttons (A20)
	Health Protection (B9)	First-aid stations (A21)
	Disaster Preparedness (B10)	Emergency plans and signage for fire, flood, and earthquake (A22)
	Surveillance and Lighting (B11)	Surveillance in key areas (A23), Adequate night lighting (A24), Safety in nighttime activities (A25)
Social and Emotional Belonging Needs (C3)	Social Interaction (B12)	Spaces for conversation (A26), Opportunities for making friends (A27)
	Activity Participation (B13)	Ancestral worship events (A28), Traditional festivals (A29), Village decision-making gatherings (A30)
	Sense of Belonging (B14)	Familiarity (A31), Sense of identity (A32), Feeling at home (A33)
	Intergenerational Integration (B15)	Shared use with younger generations, e.g., elderly caring for children (A34)
	Volunteering and Mutual Aid (B16)	Volunteer service stations (A35), Encouragement of mutual care (A36), Warm atmosphere (A37)
Environmental and Maintenance Needs (C4)	Environmental Comfort (B17)	Fresh air (A38), Green landscaping (A39), Low noise levels (A40)
	Sanitation Conditions (B18)	Cleanliness (A41), Sufficient trash bins (A42), Routine cleaning (A43)
	Management and Maintenance (B19)	On-site personnel (A44), Timely facility repair (A45), Environmental upkeep (A46)
	Operational Cost Efficiency (B20)	Low construction costs (A47), Minimal maintenance expenses (A48)
	Localized Information Boards (B21)	Community bulletin boards (A49), Informative signage and slogans (A50)
	Architectural Localization (B22)	Use of traditional local styles (A51), Integration of local culture (A52)
	Sustainable Materials (B23)	Eco-friendly materials (A53), Green-certified materials (A54), Non-toxic materials (A55)

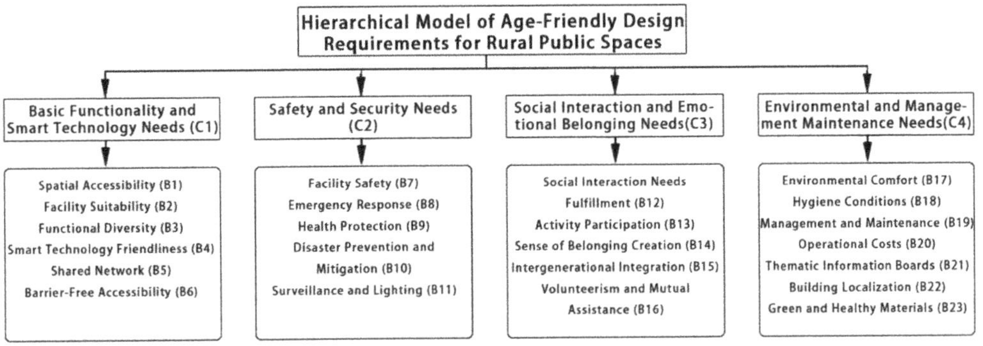

Figure 5. Hierarchical Model of Demand Elements.

(5) Theoretical Saturation Testing

To ensure completeness and robustness, a theoretical saturation test was conducted to verify that all critical categories and relationships had been adequately captured. Two additional interview transcripts, reserved for validation, were analyzed, yielding no new core categories or key indicators. This confirmed that the current coding structure had reached the point of theoretical saturation. Overall, the coding framework effectively captures the diverse and layered nature of elderly spatial needs in rural settings, offering valuable theoretical support for both policy-making and practical design application.

4.2. User Demand Screening Based on the FKANO Model

To further explore and accurately interpret the latent and affective preferences of elderly users regarding rural public space, this study employed the FKANO model to classify and analyze user needs. Based on the demand indicators established in Table 2, a structured FKANO questionnaire was developed. Each item was assessed through paired positive and negative phrasing, allowing for dual evaluation of presence versus absence. The survey was administered to 90 elderly individuals and family members, along with 20 village administrators, resulting in 100 valid responses after the exclusion of six inconsistent and four incomplete questionnaires.

The FKANO model applied a fuzzy interval scale of [0, 1] to capture the degree of importance and satisfaction associated with each spatial attribute. This enabled the fuzzy categorization and quantitative prioritization of user needs, identifying the most critical features and functions of rural public spaces from an elderly-centric perspective. The results support targeted spatial enhancements by designers and administrators, ensuring that the configuration of public spaces is more closely aligned with user expectations. This approach also contributes to improving the attractiveness and satisfaction of rural community services. To illustrate the application of the model, Table 3 presents selected items from the FKANO questionnaire used in the empirical study. The findings provide nuanced insights into user preference structures and offer actionable evidence for context-sensitive optimization of age-friendly rural environments.

Table 3. FKANO Model Survey Questionnaire (Excerpt).

Type	User Demand		Like	Must-Be	Indifferent	Tolerable	Dislike
Basic Functional and Cognitive Needs (C1)	Spatial Convenience (B1)	Present Absent	0.7 0	0.3 0	0 0.2	0 0.2	0 0.6
…	…	…	…	…			
Safety and Protection Needs (C2)	Facility Security (B7)	Present Absent	0.5 0	0.2 0	0.3 0	0 0.7	0 0.3
…	…	…	…	…	…	…	…

A total of 100 valid questionnaires were collected from rural elderly residents and village administrators. Given the volume of data, Item B1 from Table 3 was chosen as a representative example. The presence vector for this item was defined as P = [0.7, 0.3, 0, 0, 0], and the absence vector as N = [0, 0, 0.2, 0.2, 0.6]. Based on these vectors, the interaction matrix S was constructed as follows:

$$S = \begin{bmatrix} 0 & 0 & 0.14 & 0.14 & 0.42 \\ 0 & 0 & 0.06 & 0.06 & 0.18 \\ 0 & 0 & 0 & 0 & 0 \\ 0 & 0 & 0 & 0 & 0 \\ 0 & 0 & 0 & 0 & 0 \end{bmatrix}$$

By aligning the elements of Table 1 (the FKANO evaluation matrix) with the values in S, the requirement membership vector for user demand is derived as follows:

$$T_1 = \left(\frac{0.18}{M'}, \frac{0.42}{O'}, \frac{0.28}{A'}, \frac{0.12}{I'}, \frac{0}{R'} \right)$$

According to the FKANO classification rules, if multiple requirement categories in vector T have non-zero values, the final classification is determined by prioritizing the categories in the following order: M > O > A > I > R. In this case, since the highest value in T_1 corresponds to the One-dimensional category, the attribute for Item B1 is defined as: T_1 = (0, 1, 0, 0, 0).

This classification procedure is repeated for all questionnaire responses. For each user requirement, the most frequently occurring category is identified and designated as its final classification. Subsequently, based on Equations (1) and (2), the Better–Worse coefficients are computed. These coefficients represent the satisfaction gain index and the dissatisfaction reduction index, respectively. When the absolute values of these coefficients approach zero, it indicates that the corresponding requirement has minimal impact on user satisfaction. Conversely, values approaching one reflect stronger influence. The detailed results are presented in Table 4.

Based on the hierarchical model of age-friendly design needs for rural public spaces and the classification results derived from the FKANO model, the study identified eight Must-be (M) requirements, seven One-dimensional (O) requirements, six Attractive (A) requirements, and two Indifferent (I) requirements, as shown in Table 5. According to FKANO theory, Indifferent attributes exert minimal influence on satisfaction and are thus excluded from further analysis.

Table 4. FKANO Model Classification Results.

ID	Attribute Count						Classified Attribute	Better Coefficient (%)	Worse Coefficient (%)
	A	*O*	*M*	*I*	*R*	*Q*			
B1	18	12	43	15	6	6	*M*	27.3	−41.6
B2	13	44	15	18	5	5	*O*	55.6	−40.4
B3	14	41	17	20	4	4	*O*	54.1	−39.7
B4	41	18	18	13	5	5	*A*	65.1	−33.3
B5	38	15	20	17	6	4	*A*	60.0	−29.2
B6	15	13	43	19	5	5	*M*	26.7	−37.8
B7	16	14	41	19	5	5	*M*	28.1	−36.8
B8	13	46	15	17	5	4	*O*	58.6	−38.5
B9	16	42	17	15	5	5	*O*	59.7	−41.0
B10	15	43	18	15	5	4	*O*	58.7	−41.3
B11	18	13	43	16	5	5	*M*	29.4	−36.6
B12	15	14	41	20	5	5	*M*	27.7	−36.1
B13	15	43	17	16	5	4	*O*	59.3	−40.7
B14	38	17	20	16	5	4	*A*	60.6	−28.8
B15	13	44	16	18	5	4	*O*	55.7	−40.5
B16	41	14	20	16	5	4	*A*	65.1	−32.5
B17	15	13	25	39	4	4	*M*	34.0	−34.7
B18	14	12	46	18	6	4	*M*	26.0	−36.1
B19	12	12	21	42	7	6	*I*	29.2	−30.0
B20	10	11	14	55	6	4	*I*	27.6	−25.8
B21	9	11	14	57	5	4	*I*	26.0	−24.0
B22	39	13	19	20	5	4	*A*	57.4	−28.2
B23	41	12	17	22	4	4	*A*	67.7	−29.0

Table 5. Classification of Age-Friendly Design Requirements for Rural Public Spaces Based on the FKANO Model.

FKANO Category	Design Requirement Classification
M (Retained)	Spatial Accessibility (B1), Barrier-free Accessibility (B6), Facility Safety (B7), Surveillance and Lighting (B11), Social Interaction Needs Fulfillment (B12), Environmental Comfort (B17), Hygiene Conditions (B18), Management and Maintenance (B19)
O (Retained)	Facility Suitability (B2), Functional Diversity (B3), Emergency Response (B8), Health Protection (B9), Disaster Prevention and Mitigation (B10), Activity Participation (B13), Intergenerational Integration (B15)
A (Retained)	Smart Technology Friendliness (B4), Shared Network (B5), Sense of Belonging Creation (B14), Volunteerism and Mutual Assistance (B16), Building Localization (B22), Green and Healthy Materials (B23)
I (Not included)	Operational Costs (B20), Thematic Information Boards (B21)

The retained M, O, and A requirements, representing fundamental expectations, performance-based needs, and delight factors, are adopted as core evaluation dimensions for optimizing age-friendly design in rural public spaces, as shown in Figure 6. By removing non-impactful I-type attributes, the evaluation system becomes more user-centered and design-efficient, enhancing the targeted effectiveness of spatial interventions for aging populations in rural settings.

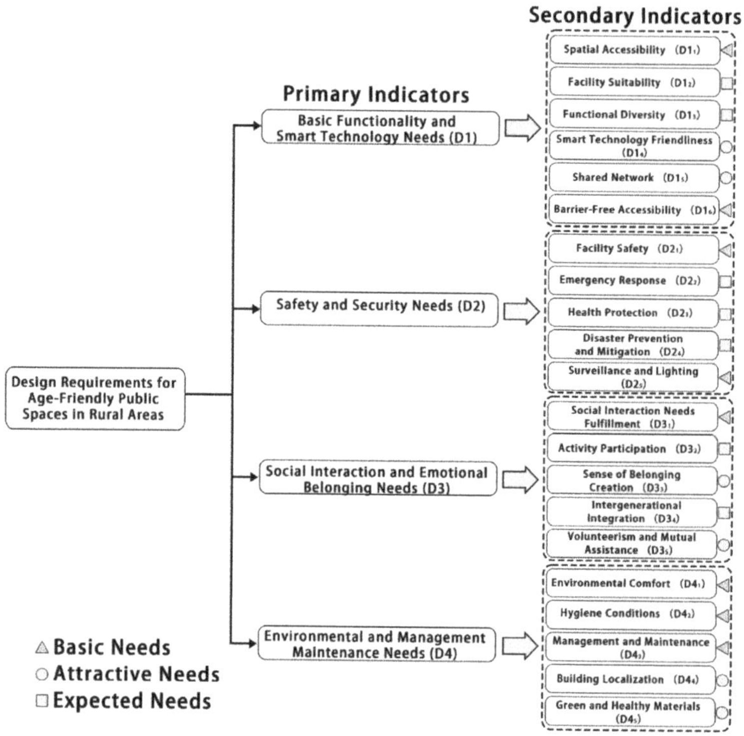

Figure 6. Hierarchical Model of Age-Friendly Design Requirements for Rural Public Spaces.

4.3. Evaluation System Construction and Weight Ranking

To further clarify the interrelationships among user needs after determining their categorical attributes, this study applies the DEMATEL (Decision-Making Trial and Evaluation Laboratory) method to analyze indicator weights. Based on the FKANO classification results, a total of 21 key indicators were extracted for age-friendly design in rural public spaces. These indicators are systematically categorized into four dimensions: Basic Functional and Intelligent Service Needs (6 indicators), Safety and Protection Needs (5 indicators), Social Interaction and Emotional Belonging Needs (5 indicators), and Environmental and Maintenance Management Needs (5 indicators). This structured framework not only reflects the multifaceted nature of elderly user demands but also provides robust input for the subsequent DEMATEL-based causal analysis, enabling a nuanced understanding of influence relationships among design criteria.

To further elucidate the interrelationships among user needs in the context of age-friendly design for rural public spaces, a DEMATEL-based expert survey was administered, involving 10 expert participants: eight university professors specializing in environmental design and two rural governance practitioners. Participants completed a structured questionnaire designed to assess the pairwise influence relationships among the 21 user needs, grouped into four major categories.

Each expert evaluated the direct influence of one need on another using a standardized scoring system, resulting in a 21×21 direct-relation matrix. As shown in Table 6, higher scores indicate a stronger direct influence of one need over another. The individual evaluations were aggregated and arithmetically averaged to generate the initial direct influence matrix, which served as the foundational dataset for the subsequent DEMATEL analysis.

Table 6. Normative Direct Influence Matrix.

	$D1_1$	$D1_2$	$D1_3$	$D1_4$	$D1_5$	$D1_6$	$D2_1$	$D2_2$	$D2_3$	$D2_4$	$D2_5$	$D3_1$	$D3_2$	$D3_3$	$D3_4$	$D3_5$	$D4_1$	$D4_2$	$D4_3$	$D4_4$	$D4_5$
$D1_1$	0.000	0.063	0.048	0.041	0.033	0.074	0.059	0.052	0.041	0.059	0.041	0.048	0.052	0.037	0.045	0.045	0.067	0.033	0.045	0.030	0.033
$D1_2$	0.074	0.000	0.056	0.063	0.026	0.063	0.056	0.030	0.037	0.048	0.041	0.059	0.048	0.041	0.052	0.048	0.056	0.030	0.041	0.033	0.030
$D1_3$	0.063	0.067	0.000	0.041	0.052	0.048	0.041	0.022	0.026	0.033	0.030	0.041	0.048	0.041	0.048	0.048	0.056	0.033	0.033	0.033	0.037
$D1_4$	0.041	0.056	0.071	0.000	0.067	0.052	0.041	0.030	0.030	0.037	0.041	0.048	0.048	0.045	0.052	0.048	0.059	0.030	0.030	0.030	0.033
$D1_5$	0.048	0.037	0.048	0.048	0.000	0.041	0.019	0.056	0.030	0.030	0.041	0.056	0.041	0.037	0.045	0.048	0.052	0.022	0.045	0.022	0.019
$D1_6$	0.059	0.048	0.056	0.045	0.026	0.000	0.056	0.056	0.030	0.045	0.037	0.063	0.063	0.052	0.045	0.045	0.059	0.022	0.045	0.030	0.037
$D2_1$	0.052	0.067	0.041	0.048	0.019	0.063	0.000	0.063	0.048	0.071	0.048	0.056	0.059	0.048	0.045	0.048	0.059	0.030	0.059	0.030	0.041
$D2_2$	0.059	0.048	0.037	0.041	0.056	0.063	0.063	0.000	0.048	0.074	0.056	0.030	0.041	0.037	0.026	0.033	0.037	0.033	0.045	0.026	0.026
$D2_3$	0.045	0.052	0.041	0.033	0.030	0.052	0.045	0.045	0.000	0.052	0.033	0.033	0.041	0.045	0.045	0.037	0.048	0.045	0.037	0.030	0.041
$D2_4$	0.056	0.037	0.037	0.033	0.041	0.048	0.059	0.067	0.048	0.000	0.052	0.033	0.037	0.033	0.022	0.048	0.045	0.041	0.041	0.022	0.041
$D2_5$	0.056	0.041	0.026	0.030	0.045	0.037	0.048	0.063	0.052	0.052	0.000	0.045	0.056	0.048	0.033	0.026	0.059	0.037	0.052	0.037	0.030
$D3_1$	0.037	0.048	0.045	0.048	0.052	0.045	0.026	0.022	0.022	0.022	0.033	0.000	0.067	0.071	0.052	0.045	0.059	0.037	0.045	0.048	0.026
$D3_2$	0.048	0.033	0.045	0.037	0.033	0.037	0.048	0.030	0.033	0.022	0.033	0.059	0.000	0.063	0.048	0.045	0.052	0.026	0.037	0.037	0.030
$D3_3$	0.033	0.045	0.037	0.041	0.019	0.048	0.037	0.030	0.030	0.026	0.033	0.056	0.059	0.000	0.045	0.052	0.041	0.041	0.052	0.048	0.052
$D3_4$	0.041	0.041	0.041	0.056	0.030	0.045	0.052	0.026	0.048	0.037	0.033	0.056	0.052	0.056	0.000	0.041	0.045	0.030	0.033	0.048	0.033
$D3_5$	0.033	0.030	0.033	0.041	0.026	0.041	0.037	0.030	0.033	0.026	0.022	0.052	0.048	0.048	0.048	0.000	0.041	0.041	0.037	0.030	0.026
$D4_1$	0.063	0.056	0.045	0.045	0.037	0.059	0.048	0.033	0.052	0.037	0.045	0.063	0.059	0.056	0.045	0.045	0.000	0.063	0.052	0.041	0.056
$D4_2$	0.056	0.041	0.037	0.033	0.019	0.045	0.041	0.033	0.045	0.041	0.022	0.052	0.048	0.052	0.045	0.045	0.059	0.000	0.048	0.030	0.045
$D4_3$	0.052	0.026	0.041	0.056	0.030	0.052	0.059	0.063	0.041	0.056	0.056	0.041	0.045	0.045	0.033	0.037	0.056	0.048	0.000	0.030	0.033
$D4_4$	0.026	0.026	0.022	0.019	0.019	0.041	0.019	0.026	0.022	0.022	0.026	0.033	0.030	0.059	0.045	0.030	0.056	0.026	0.026	0.000	0.033
$D4_5$	0.045	0.056	0.037	0.033	0.019	0.048	0.041	0.022	0.045	0.026	0.022	0.033	0.030	0.045	0.037	0.019	0.056	0.045	0.033	0.026	0.000

After deriving the total influence matrix (T), the influence degree (D) and dependence degree (R) for each user need were calculated. Influence (D) quantifies the extent to which a given need affects others within the system, while dependence (R) reflects the degree to which it is affected by others. The sum (D + R), referred to as prominence or centrality, indicates the systemic importance of each need within the overall interaction network. The difference (D − R), known as the causal degree, distinguishes causal factors (positive values) from effect factors (negative values). These indicator values are presented in Table 7. A positive causal degree identifies the need as an active influencer, whereas a negative value denotes a passive receiver of influence. This analysis provides critical insights into the structural dynamics and interdependencies among user needs, informing the prioritization of design interventions in age-friendly rural public spaces.

Table 7. DEMATEL Computed Indicator Values and Weights.

	Influence Degree (D)	Dependence Degree (C)	Prominence (D + C)	Causal Degree (D − C)	Weight
$D1_1$	6.651	6.863	13.514	−0.212	0.0542
$D1_2$	6.538	6.414	12.952	0.124	0.0519
$D1_3$	5.904	5.929	11.832	−0.025	0.0474
$D1_4$	6.187	5.855	12.042	0.332	0.0483
$D1_5$	5.508	4.758	10.266	0.750	0.0411
$D1_6$	6.426	6.977	13.404	−0.551	0.0537
$D2_1$	6.955	6.280	13.235	0.675	0.0530
$D2_2$	6.197	5.551	11.748	0.645	0.0471
$D2_3$	5.808	5.310	11.118	0.498	0.0446
$D2_4$	5.929	5.698	11.627	0.230	0.0466
$D2_5$	6.118	5.252	11.370	0.866	0.0456
$D3_1$	5.893	6.724	12.617	−0.830	0.0506
$D3_2$	5.585	6.835	12.420	−1.249	0.0498
$D3_3$	5.708	6.665	12.372	−0.957	0.0496
$D3_4$	5.864	5.974	11.839	−0.110	0.0474
$D3_5$	5.078	5.854	10.932	−0.776	0.0438
$D4_1$	6.918	7.322	14.239	−0.404	0.0571
$D4_2$	5.852	5.012	10.864	0.840	0.0435
$D4_3$	6.310	5.855	12.166	0.455	0.0487
$D4_4$	4.260	4.694	8.954	−0.433	0.0359
$D4_5$	5.090	4.957	10.047	0.133	0.0403

By plotting prominence (D + R) and causal degree (D − R) on the horizontal and vertical axes, respectively, an Influence Network Relationship Map (INRM) was constructed to visualize the causal structure of user needs in age-friendly design for rural public spaces. In this network, the direction of arrows represents the flow of influence—from active, causal needs toward those more passively affected, allowing designers and planners to intuitively identify critical needs and their transmission pathways.

The prominence value reflects the systemic significance of each factor; the higher the value, the more central the need is within the network. Accordingly, the INRM (see Figure 7) provides a graphical representation of the causal and influence dynamics among all 21 indicators. Based on this analysis, the hierarchical importance of user needs in rural age-friendly design is determined as follows: D41 > D11 > D16 > D21 > D31 > D12 > D32 > D33 > D43 > D14 > D22 > D24 > D13 > D34 > D23 > D35 > D42 > D15 > D45 > D25 > D44.

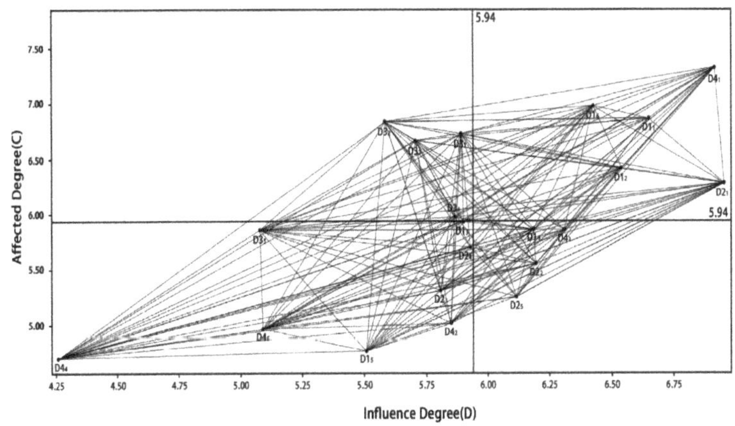

Figure 7. Influence–Dependence Map.

This ranking highlights the most influential needs within the system, offering a data-driven foundation for prioritizing design interventions and policy actions that address the challenges of aging in rural environments.

4.4. Age-Friendly Design Practices for Rural Public Spaces

The core objective of this study is to optimize design decision-making for age-friendly rural public spaces by anchoring the process in user–designer cognitive synergy. Grounded Theory was used to elicit and structure the multidimensional cognitions of older users and designers. Needs were then classified with the Fuzzy Kano (FKANO) model, and the DEMATEL method mapped causal relations to identify priority drivers. Results highlighted Environmental and Operations Management (D41, Must-be; highest prominence) and Spatial Accessibility (D11, Must-be) as critical determinants of satisfaction and usability, while Smart-Friendliness (D16, Attractive) and Safety Organization (D21, Must-be) played important roles in enhancing gains and reducing dissatisfaction.

(1) Overall Spatial Layout Design

The layout adopts multifunctional zoning and separates dynamic (recreation, play, fitness) from static (reading, rest) areas (Figure 8). All zones are linked by continuous, barrier-free pathways, addressing Spatial Accessibility (D11, M) and reducing resistance for mobility aids. Glass curtain walls mitigate noise transfer, improving rest quality and fulfilling Environmental Comfort (B17, M).

(2) Design of Key Spaces and Facilities

Barrier-free accessibility and safety features are emphasized. Pathways use low-gradient slopes and anti-slip surfaces; high-visibility signage assists orientation, meeting Barrier-free Accessibility (B6, M) and Safety Organization (D21, M). Restrooms are ergonomically adapted for elderly needs, as shown in Figure 9. Smart technologies, such as motion-sensor lighting, public Wi-Fi, and one-touch emergency call points, address Smart-Friendliness (D16, A) and Shared Network (B5, A). Service desks integrate first-aid kits, AEDs, and surveillance, operationalizing Emergency Response (B8, O) and Health Protection (B9, O), as shown in Figure 10.

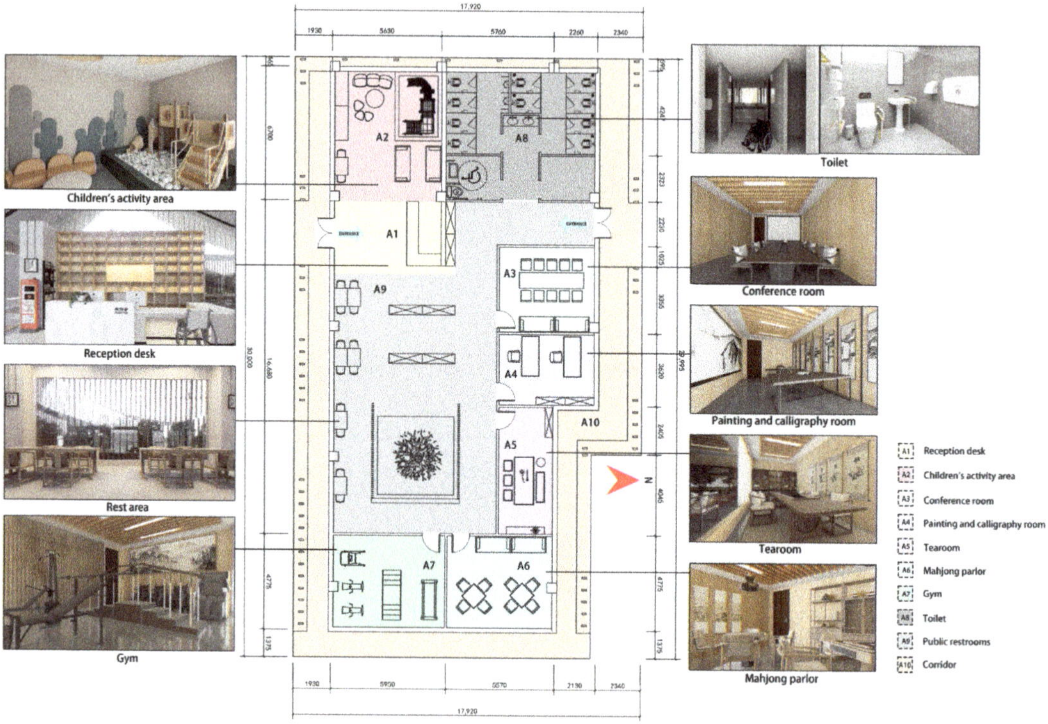

Figure 8. Interior Floor Plan.

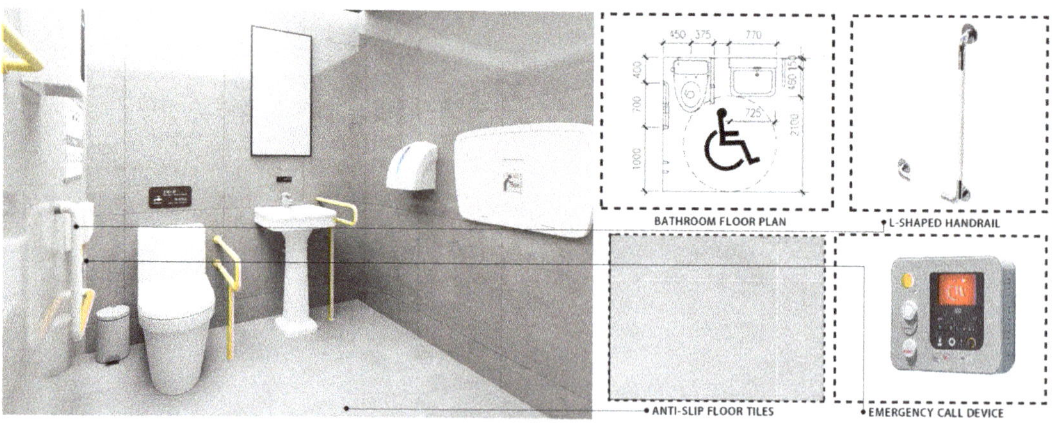

Figure 9. Barrier-free Bathroom Design.

(3) Creation of Social and Emotional Spaces

Interactive spaces such as chess rooms, tea lounges, and multipurpose areas (Figure 11) foster community communication. Intergenerational zones combine children's play areas with nearby seating for grandparents, implementing Social Interaction (D31, M) and Activity Participation (B13, O). Modular furniture and flexible layouts enhance inclusivity and embed Sense of Belonging (B14, A) and Intergenerational Integration (B15, O).

Figure 10. Service Desk Design.

Figure 11. Communication and Interaction Space.

(4) Design Highlights and Innovations

Key user needs were transformed into concrete design strategies. An operations hub consolidates maintenance and custodial functions, addressing Environmental and Operations Management (D41, M) and Hygiene Conditions (B18, M). Smart services and local cultural symbols (B22, A; B23, A) enrich usability and identity while avoiding thematic overload. Compared with designs lacking such analytical grounding, this scheme ensures continuity in accessibility, robust safety infrastructure, and sustainable upkeep, achieving a systematic alignment between user priorities and spatial interventions.

5. Discussion

To evaluate the practical feasibility and user satisfaction of the optimized age-friendly rural public space design, a comprehensive post-occupancy evaluation was conducted. The assessment involved 20 elderly users and 5 grassroots government officials, who rated the design across five critical dimensions: (1) safety and accessibility, (2) functional suitability and convenience, (3) support for social participation and interaction, (4) environmental comfort, and (5) cultural identity and community sustainability.

Each criterion was evaluated using a five-point Likert scale, ranging from 1 ("very dissatisfied") to 5 ("very satisfied"). The collected responses were subjected to statistical analysis to calculate the mean score for each evaluation dimension. The results, visually presented in a radar chart (see Figure 12), reveal a consistently high level of satisfaction, with average scores exceeding 4.0 across all categories. These findings provide strong empirical support for the real-world applicability and user-centered effectiveness of the proposed design.

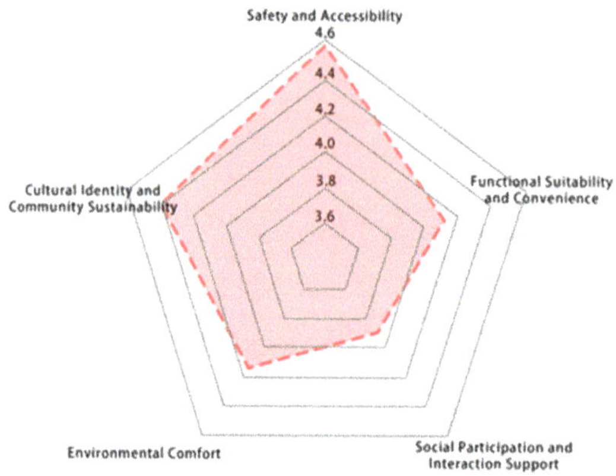

Figure 12. User Evaluation Results.

The uniformly positive feedback from both elderly participants and local officials not only validates the core design principles but also confirms their alignment with actual user needs and community priorities. Moreover, the evaluation offers valuable insights for future improvements, serving as a data-driven foundation for scaling and replicating age-friendly spatial strategies in other rural contexts. Finally, the results underscore the practical value of participatory design, highlighting the role of multi-stakeholder engagement in the iterative development of inclusive public-space interventions.

In addition to immediate feasibility, long-term adaptability across diverse rural settings must be considered. Rural communities differ significantly in geography, climate, infrastructure, and socio-cultural traditions, which may influence the prioritization of user needs and the effectiveness of design strategies. For example, mountainous villages require enhanced anti-slip measures and slope management, while coastal or plain areas may emphasize shading, ventilation, and flood resilience. To ensure sustainability, the proposed framework incorporates an operations hub and continuous post-occupancy evaluation, allowing local stakeholders to adjust facility maintenance, safety standards, and social programming over time. This iterative mechanism not only strengthens long-term usability and cost-effectiveness but also supports context-specific adaptation, making the framework scalable to varied rural environments.

6. Conclusions

This study developed a User–Designer Cognitive Alignment Framework to enhance the age-friendliness of rural public spaces. By integrating Grounded Theory, the Fuzzy Kano model, and the DEMATEL method, the research systematically identified, classified, and weighted the multi-dimensional needs of older adults, bridging the cognitive gap between users and designers. The case application in Jiangsu Province demonstrated that the framework can translate complex needs into actionable spatial strategies, such as continuous barrier-free circulation, robust nighttime safety systems, and integrated operations hubs, thereby improving accessibility, safety, social belonging, and environmental sustainability. These findings contribute both theoretical enrichment and practical guidance for participatory and evidence-based rural design.

Nonetheless, this study has limitations. The empirical testing was restricted to a small number of villages in Jiangsu Province, which constrains the generalizability of the findings. Regional variations in socio-cultural traditions, geographical conditions, and infrastructure levels may influence user needs and design priorities. In addition, the sample size and focus on relatively well-developed rural areas may underrepresent challenges in less-resourced or climatically extreme settings. These factors should be considered when transferring the model to other contexts.

Future research will therefore focus on enhancing robustness and scalability. First, expanding the study to diverse rural regions (e.g., North China Plain, Southwest mountainous areas, Southeast coastal zones) will allow comparative analysis of cultural and geographical differences. Second, integrating climate and cultural factors as moderating variables into the FKANO–DEMATEL framework can reveal regional variations in need prioritization [44]. Third, establishing a demand–design feedback loop through continuous post-occupancy evaluation will enable dynamic recalibration of design standards. Together, these steps will improve the adaptability and replicability of the proposed framework, supporting its broader application in rural aging-friendly design.

Author Contributions: Conceptualization, Z.Z. (Zhihuan Zhang); methodology, Z.Z. (Zhihuan Zhang); software, Z.Z. (Zhihuan Zhang); validation, Z.Z. (Zhihuan Zhang); formal analysis, Z.Z. (Zhihuan Zhang); investigation, Z.Z. (Ziqi Zhan); resources, Z.Z. (Zhihuan Zhang); data curation, Z.Z. (Ziqi Zhan); writing—original draft preparation, Z.Z. (Zhihuan Zhang); writing—review and editing, Z.Z. (Zhihuan Zhang); visualization, Z.Z. (Zhihuan Zhang); supervision, Y.L.; project administration, Y.L.; funding acquisition, Y.L. All authors have read and agreed to the published version of the manuscript.

Funding: This study received no external funding support.

Institutional Review Board Statement: The study only involves a minimal-risk, general questionnaire survey. It does not involve any experimental procedures, sensitive personal information, or interventions with participants. Therefore, ethical approval from an Institutional Review Board was not required for this research.

Informed Consent Statement: While we did not sign a formal written consent form during the survey process, we strictly ensured that all participants verbally agreed to participate after we clearly explained the purpose, content, and non-sensitive nature of the survey. No participant expressed hesitation or refusal to take part.

Data Availability Statement: The original contributions presented in the study are included in the article, further inquiries can be directed to the corresponding author.

Conflicts of Interest: The authors declare no conflict of interest.

Appendix A

Table A1. Semi-Structured Interview Guide.

1. Usage of Public Spaces
1.1 Which public spaces in the village do you visit most often? What makes you choose these particular places? 1.2 On average, how much time do you spend in these spaces each day (or week)? What types of activities do you usually engage in? 1.3 What aspects of these spaces do you like the most? Are there any specific facilities or services that you are particularly satisfied with? 1.4 During adverse weather conditions (e.g., extreme heat, rain, or snow), do you still visit these public spaces? Why or why not?
2. Functional and Facility Needs
2.1 What facilities or functions do you think are still missing in the village's public spaces? Why do you think they are needed? 2.2 Which facilities are the most important to you? Why? Are there any that you find especially useful? 2.3 Are there any facilities that you feel are rarely used or unsuitable for older adults? Could you give some examples? 2.4 How satisfied are you with the existing facilities? Are there any you think should be improved or upgraded?

Table A1. *Cont.*

3. Emotions and Sense of Belonging
3.1 Do you meet or socialize with friends or neighbors in these public spaces? Is this kind of social interaction important to you? 3.2 How do these spaces influence your daily life, social relationships, or emotional well-being? Is there a particular space that makes you feel especially warm or included? 3.3 What aspects of space design make you feel a stronger sense of belonging or make you more willing to stay? Why?
4. Safety and Accessibility
4.1 Do you have any safety concerns when using public spaces? For example, fear of falling, traffic risks, etc.? Are there any areas you find particularly unsafe? 4.2 Are the spaces accessible for mobility aids such as wheelchairs or walkers? Are there specific places that are unfriendly to wheelchair users? 4.3 What are your suggestions or expectations regarding the safety and accessibility of public spaces?
5. Design Suggestions and Expectations
5.1 If public spaces could be renovated or newly built, what aspects would you most like to improve or add? Why? 5.2 Do you feel that designers truly understand your needs? Why or why not? How do you think designers could better understand your needs? 5.3 Would you like to participate in discussions or decision-making about the design of village public spaces? Why? In what ways would you prefer to be involved?

References

1. Han, S. Design of urban and rural elderly care public areas integrating person-environment fit theory. *Nonlinear Eng.—Model. Appl.* **2025**, *14*, 20250096. [CrossRef]
2. Feng, Z. *Research on the Design of Public Space of Suburban Rural Community from the Perspective of Mutual Support for the Aged: The Case of Shanghai Jinshan District*; University of Hawaii at Manoa: Honolulu, HI, USA, 2022.
3. Ren, H.G.; Yang, F.; Zhang, J.; Wang, Q. Evaluation of Cognition of Rural Public Space Based on Eye Tracking Analysis. *Buildings* **2024**, *14*, 1525. [CrossRef]
4. Nutt, N.; Salmistu, S.; Kupper, K.; Kotval, Z. Assessing age-friendliness of contemporary urban outdoor places in Estonia. *Qual. Ageing Older Adults* **2024**, *25*, 204–219. [CrossRef]
5. Hartt, M.; Gryfe, M. Age-friendly planning and play: The Study of Older Adult Play Spaces (SOAPS) observation tool. *Plan. Pract. Res.* **2025**, *40*, 24–40. [CrossRef]
6. Gong, M.; Wang, N.; Chu, Y.; Wu, Y.; Huang, J.; Wu, J. Optimizing Public Spaces for Age-Friendly Living: Renovation Strategies for 1980s Residential Communities in Hangzhou, China. *Buildings* **2025**, *15*, 211. [CrossRef]
7. Zeng, Y.J.; Chen, B. Evaluation of Aging-Friendly Public Spaces in Old Urban Communities Based on IPA Method-A Case Study of Shouyi Community in Wuhan. *Buildings* **2024**, *14*, 2362. [CrossRef]
8. Yi, C.; Han, W.; Liu, Y.; Lin, Y.; Qi, Y. Aging Adaptation Transition of Health Care Buildings for Accessibility Optimization for the Elderly. *Buildings* **2025**, *15*, 379. [CrossRef]
9. Cao, Y.; Li, D.; Gao, X.; Bi, S.; Yu, K.; Zhou, D. Age-Friendly Environment Design of High-Speed Railway Stations from a Healthy Ageing Perspective: A Case Implementation in Nanjing, China. *Buildings* **2024**, *14*, 3280. [CrossRef]
10. Hancock, S.; Winterton, R.; Wilding, C.; Blackberry, I. Understanding ageing well in Australian rural and regional settings: Applying an age-friendly lens. *Aust. J. Rural. Health* **2019**, *27*, 298–303. [CrossRef] [PubMed]
11. Helbich, M.; Yao, Y.; Liu, Y.; Zhang, J.; Liu, P.; Wang, R. Using deep learning to examine street view green and blue spaces and their associations with geriatric depression in Beijing, China. *Environ. Int.* **2019**, *126*, 107–117. [CrossRef]
12. Lin, J.J.; Li, X.Y.; Lin, J.H. Evaluation of Age-Appropriate Public Seats in Comprehensive Parks and Sustainable Design Strategies Based on the Kano-Importance-Performance Analysis Model. *Sustainability* **2024**, *16*, 6914. [CrossRef]
13. Zhang, Z.L.; Tang, X.M.; Wang, Y. Evaluation of the Intergenerational Equity of Public Open Space in Old Communities: A Case Study of Caoyang New Village in Shanghai. *Land* **2023**, *12*, 1347. [CrossRef]
14. Jiang, J.Y.; Xia, Z.; Sun, X.; Wang, X.; Luo, S. Social Infrastructure and Street Networks as Critical Infrastructure for Aging Friendly Community Design: Mediating the Effect of Physical Activity. *Int. J. Environ. Res. Public Health* **2022**, *19*, 11842. [CrossRef] [PubMed]
15. Zhou, M.; Zhu, N.; Peng, X.; Huang, J.; Ji, L.; Meng, X.; Song, S.; Liu, S. Evaluation Study of Public Interaction Spaces for the Elderly in a Community-Taking the Railway Station Community in Jiaozuo City as an Example. *Buildings* **2025**, *15*, 517. [CrossRef]

16. Han, J.W.; Ma, H.; Wang, M.; Li, J. Construction and improvement strategies of an age-friendly evaluation system for public spaces in affordable housing communities: A case study of Shenzhen. *Front. Public Health* **2024**, *12*, 1399852. [CrossRef]
17. Gong, M.; Ren, M.; Dai, Q.; Luo, X. Aging-Suitability of Urban Waterfront Open Spaces in Gongchen Bridge Section of the Grand Canal. *Sustainability* **2019**, *11*, 6095. [CrossRef]
18. Han, B.; Yu, J. Research on a novel evaluation criterion of rural public space environmental capacity based on ahp fuzzy method. *Fresenius Environ. Bull.* **2020**, *29*, 3818–3826.
19. Witteman, H.O.; Vaisson, G.; Provencher, T.; Dansokho, S.C.; Colquhoun, H.; Dugas, M.; Fagerlin, A.; MC Giguere, A.; Haslett, L.; Hoffman, A.; et al. An 11-Item Measure of User- and Human-Centered Design for Personal Health Tools (UCD-11): Development and Validation. *J. Med. Internet Res.* **2021**, *23*, e15032. [CrossRef]
20. Wang, T.; Zhao, Y.; Zhang, X.; Xie, Y.; Pang, L.L.L. Design of walking aids for the elderly based on the Kano-AHP-FEC method. *Sci. Rep.* **2025**, *15*, 2663. [CrossRef]
21. Du, S. Hybrid Kano-DEMATEL-TOPSIS model based benefit distribution of multiple logistics service providers considering consumer service evaluation of segmented task. *Expert Syst. Appl.* **2023**, *213*, 119292. [CrossRef]
22. Liu, H.; Zhu, J.; Ni, P.; Li, Y.; Li, S. Research on Design Methods for Interactive Spaces in Schools for Children with Intellectual Disabilities Considering User Needs. *Buildings* **2024**, *14*, 2230. [CrossRef]
23. Liu, X.X. Research on Function Optimization of Electric Vehicle Charging Stations Based on User Demand Analysis: An Empirical Study Using the Kano Model and AHP Method. *Sustainability* **2024**, *16*, 10783. [CrossRef]
24. Qin, M.Q.; Wang, J.; Ding, X.; Zhang, H. Symmetrical Traditional Patterns and User Perception: A Study on Innovation in Home Textile Design. *Symmetry* **2025**, *17*, 960. [CrossRef]
25. Wang, J.J.; Liang, Q.; Ma, X.; Wei, Y.; Chen, Y. Research on the Design of Growable Solid Wood Children's Beds. *Bioresources* **2024**, *19*, 8257–8272. [CrossRef]
26. Wang, L.; Gong, Y.-H.; Niu, S.-F.; Ren, Z.-X. Cultural Symbols Migration and User Perception: Innovation in Chinese-Style Furniture Design. *Bioresources* **2025**, *20*, 3476–3496. [CrossRef]
27. Chun Tie, Y.; Birks, M.; Francis, K. Grounded theory research: A design framework for novice researchers. *SAGE Open Med.* **2019**, *7*, 2050312118822927. [CrossRef]
28. Turner, C.; Astin, F. Grounded theory: Whatmakes a grounded theory study? *Eur. J. Cardiovasc. Nurs.* **2021**, *20*, 285–289. [CrossRef]
29. Foley, G.; Timonen, V. Using Grounded Theory Method to Capture and Analyze Health Care Experiences. *Health Serv. Res.* **2015**, *50*, 1195–1210. [CrossRef] [PubMed]
30. Kushner, K.E.; Morrow, R. Grounded theory, feminist theory, critical theory—Toward theoretical triangulation. *Adv. Nurs. Sci.* **2003**, *26*, 30–43. [CrossRef]
31. Zhang, M.; Pan, Y. Design of Sustainable Senior-Friendly Community Transportation Services. *Sustainability* **2021**, *13*, 13078. [CrossRef]
32. Wen, Z.; Yang, J.; Sun, B.; Liu, Y. Research on Symmetry Optimization of Designer Requirements and Prototyping Platform Functionality in the Context of Agile Development. *Symmetry* **2025**, *17*, 502. [CrossRef]
33. Li, X.; Lin, J.; Li, Q.; Cao, W. Research on assist design strategy of electric garden tools driven by data and intelligence. *Adv. Des. Res.* **2024**, *2*, 55–62. [CrossRef]
34. Zhang, F.; Li, X.; Xu, S.; Zhao, X.; Gao, Y. Ergonomic design of safety protective wearables in confined space operations. *Adv. Des. Res.* **2024**, *2*, 137–150. [CrossRef]
35. Chen, C.C.; Chuang, M.C. Integrating the Kano model into a robust design approach to enhance customer satisfaction with product design. *Int. J. Prod. Econ.* **2008**, *114*, 667–681. [CrossRef]
36. Jiang, X.Q.; Zhang, J.; Yang, C.; Wan, R. Evaluating the service quality of insular and coastal recreational fisheries by integration of the SERVQUAL-fuzzy Kano model and importance-performance analysis. *Ocean. Coast. Manag.* **2023**, *243*, 106753. [CrossRef]
37. Uzun, I.M.; Cebi, S. A novel approach for classification of occupational health and safety measures based on their effectiveness by using fuzzy Kano model. *J. Intell. Fuzzy Syst.* **2020**, *38*, 589–600. [CrossRef]
38. Wu, M.; Wang, L. A continuous fuzzy Kano's model for customer requirements analysis in product development. *Proc. Inst. Mech. Eng. Part B-J. Eng. Manuf.* **2012**, *226*, 535–546. [CrossRef]
39. Si, S.L.; You, X.-Y.; Liu, H.-C.; Zhang, P. DEMATEL Technique: A Systematic Review of the State-of-the-Art Literature on Methodologies and Applications. *Math. Probl. Eng.* **2018**, *2018*, 3696457. [CrossRef]
40. Lin, C.J.; Wu, W.W. A causal analytical method for group decision-making under fuzzy environment. *Expert Syst. Appl.* **2008**, *34*, 205–213. [CrossRef]
41. Yang, J.; Yin, X.; Xu, Y.; Gu, Z.; Li, T.; Yan, Y.; Jiang, Q.; Lu, T.; Fu, Y.; Yu, Y.; et al. Medication safety analysis of elderly inpatients based on improved functional resonance analysis method (FRAM): A mixed methods study. *Sci. Rep.* **2025**, *15*, 19496. [CrossRef] [PubMed]

42. Wu, C.H. An Empirical Study on Selection, Evaluation, and Management Strategies of Green Suppliers in Manufacturing Enterprises. *J. Organ. End User Comput.* **2022**, *34*, 1–18. [CrossRef]
43. Yang, Y.; Shieh, H.M.; Tzeng, G.H. A VIKOR technique based on DEMATEL and ANP for information security risk control assessment. *Inf. Sci.* **2013**, *232*, 482–500. [CrossRef]
44. Huo, K.P.; Qin, R.; Zhao, J.; Ma, X. Long-term tracking of urban structure and analysis of its impact on urban heat stress: A case study of Xi'an, China. *Ecol. Indic.* **2025**, *174*, 113418. [CrossRef]

buildings

MDPI

Article

Evaluating the Microclimatic Performance of Elevated Open Spaces for Outdoor Thermal Comfort in Cold Climate Zones

Xuan Ma [1], Qian Luo [1], Fangxi Yan [1], Yibo Lei [1], Yuyang Lu [1], Haoyang Chen [1], Yuhuan Yang [1], Han Feng [2], Mengyuan Zhou [3], Hua Ding [1] and Jingyuan Zhao [1,*]

1 Department of Architecture, Chang'an University, Xi'an 710064, China; mxozil@yahoo.com (X.M.); 2021900773@chd.edu.cn (Q.L.); 2021900633@chd.edu.cn (F.Y.); 2021904855@chd.edu.cn (Y.L.); 2018901992@chd.edu.cn (Y.L.); 2018901199@chd.edu.cn (H.C.); 18211960609@163.com (Y.Y.); dinghua@chd.edu.cn (H.D.)
2 Department of Art, Shaanxi Normal University, Xi'an 710062, China; fenghan@snnu.edu.cn
3 Department of Architecture, The University of Kitakyushu, Fukuoka 808-0135, Japan; e3dbb410@eng.kitakyu-u.ac.jp
* Correspondence: zjyqtt@163.com

Abstract: Improving outdoor thermal comfort is a critical objective in urban design, particularly in densely built urban environments. Elevated semi-open spaces—outdoor areas located beneath raised building structures—have been recognized for enhancing pedestrian comfort by improving airflow and shading. However, previous studies primarily focused on warm or temperate climates, leaving a significant research gap regarding their thermal performance in cold climate zones characterized by extreme seasonal variations. Specifically, few studies have investigated how these spaces perform under conditions typical of northern Chinese cities like Xi'an, which is explicitly classified within the Cold Climate Zone according to China's national standard GB 50176-2016 and experiences both severe summer heat and cold winter conditions. To address this gap, we conducted field measurements and numerical simulations using the ENVI-met model (v5.0) to systematically evaluate the microclimatic performance of elevated ground-floor spaces in Xi'an. Key microclimatic parameters—including air temperature, mean radiant temperature, relative humidity, and wind velocity—were assessed during representative summer and winter conditions. Our findings indicate that the height of the elevated structure significantly affects outdoor thermal comfort, identifying an optimal elevated height range of 3.6–4.3 m to effectively balance summer cooling and winter sheltering needs. These results provide valuable design guidance for architects and planners aiming to enhance outdoor thermal environments in cold climate regions facing distinct seasonal extremes.

Keywords: elevated design beneath building; on-site measurement; numerical simulation; thermal comfort

1. Introduction

Rapid urbanization has led to a substantial increase in energy consumption across many cities, contributing to the degradation of the urban climate and environment—particularly in developing countries [1]. Among the resulting challenges, the urban heat island (UHI) effect—characterized by elevated surface and air temperatures in urban areas compared to their rural counterparts—has attracted significant research attention [2]. The UHI effect not only exacerbates thermal discomfort but also intensifies energy demand for indoor cooling during hot periods [3]. For instance, studies indicate that energy consumption in urban areas is approximately 81% higher than in rural regions [4], underscoring the urgent need for effective UHI mitigation strategies.

As urban residents increasingly engage in outdoor activities such as cycling, jogging, and walking, the quality of the outdoor thermal environment has become a key factor influencing urban livability. A thermally comfortable outdoor setting encourages longer periods of outdoor stay, potentially reducing dependence on mechanical cooling and lowering overall indoor energy use.

Outdoor thermal comfort is commonly defined as "the condition of mind that expresses satisfaction with the outdoor thermal environment" [5]. Since the 1980s, interest in outdoor thermal comfort has grown, especially concerning pedestrian experiences in urban plazas, squares, and street canyons. Thermal comfort in outdoor settings is governed by the principles of human thermo-physiology and the body's heat balance with its surroundings. This exchange of heat—via radiation, evaporation, conduction, and convection—is influenced by both human-related factors (e.g., activity level, clothing insulation) and environmental parameters that shape the thermal context (Figure 1) [6].

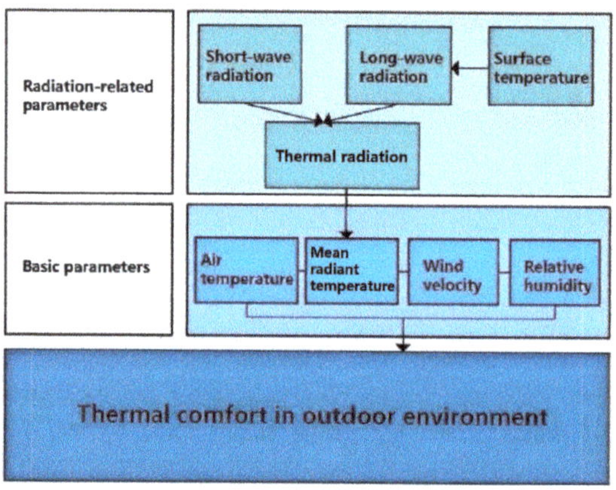

Figure 1. The parameters affecting outdoor thermal comfort.

Vital physical parameters—such as air temperature, mean radiant temperature, wind velocity, and relative humidity—are interrelated and form the basis for calculating outdoor thermal comfort.

Numerous indices have been developed to quantify thermal comfort; over 165 indicators have been identified in previous research [7]. Among them, widely recognized indices include the Standard Effective Temperature (SET*) [8] and Predicted Mean Vote (PMV) [9]. However, these two indices were originally developed for indoor environments and are generally not suitable for outdoor applications due to the dynamic and complex nature of outdoor climates. While the PMV index may have limited outdoor applicability as an indicator for the transition from thermal discomfort to physiological heat stress.

The Universal Thermal Climate Index (UTCI) [10] and Physiological Equivalent Temperature (PET) [11] are examples of such indices. In this study, PET—derived from the Munich Energy Balance Model for Individuals (MEMI) [12]—was adopted due to its strong suitability and established validity in outdoor thermal comfort research.

Extensive research has been conducted on strategies to enhance outdoor thermal comfort, focusing on factors such as urban canyon geometry and orientation [2,13–15], vegetation coverage [16], building density [17], and surface thermal properties of pavements and façades [18–20]. For instance, increasing the aspect ratio (H/W) of urban canyons has been shown to reduce maximum daytime thermal stress by up to 4 °C PET [21], while the introduction of vegetation, such as trees, may result in a PET reduction of approximately 3 °C during daytime hours [22]. Likewise, materials with high albedo for pavements and low albedo for vertical surfaces have been found to improve local thermal conditions [23].

However, most of these studies have been concentrated in warm European climates [24–27], and only more recently has research emerged in various climate zones across China, including hot-summer and cold-winter climates [28–30], hot-summer and warm-winter regions [31–35], and severely cold areas [36–38]. Despite these developments, studies specifically addressing thermal comfort in cold climate zones as defined by the Chinese national standard GB 50176-2016 remain limited [39], even though such regions experience intense heat stress during the summer (Figure 2). Moreover, existing studies have primarily focused on traditional outdoor environments such as parks [40–42], streets [1], squares [43,44], and campuses [45], while research on the microclimatic performance of semi-open areas beneath buildings is still scarce.

Figure 2. The climate zone of China (Xi'an is explicitly classified within the Cold Climate Zone according to GB 50176-2016) [39].

One architectural approach that creates such semi-open spaces is the "elevated design," which places part of a building above ground level on stilts or columns, generating shaded public areas below. Commonly seen in high-density cities like Hong Kong, this design provides sheltered outdoor space that can mitigate solar exposure on hot days. Studies have shown that wind velocity tends to be higher in elevated open spaces [46], and thermal simulations have demonstrated that these areas can exhibit PET reductions of up to 15 °C compared to adjacent open areas [47]. However, previous investigations of elevated buildings have primarily been conducted in hot-summer and warm-winter regions of southern China [46–50].

Xi'an—a city explicitly classified within the Cold Climate Zone according to GB 50176-2016—is experiencing increasingly frequent and intense summer heatwaves, becoming one of China's "new top ten hottest furnaces." Although Xi'an lies within the Cold Climate Zone, the city faces distinct seasonal extremes, including notably hot summers and cold winters. Thus, urban design strategies for Xi'an must concurrently address both summer heat stress and winter thermal discomfort to create genuinely climate-responsive urban environments. In this study, we aim to evaluate the thermal performance of elevated open spaces in Xi'an and identify optimal elevation parameters through a combination of on-site measurements and ENVI-met simulations. The findings are intended to inform design guidelines for elevated educational buildings and other semi-open structures in cold climate zones experiencing pronounced seasonal extremes, similar to Xi'an.

2. Materials and Methods

2.1. Climate Conditions of the Research City in China

Xi'an is geographically situated between 107.4° E and 109.49° E longitude and 33.42° N and 34.45° N latitude. According to the Thermal Design Code for Civil Building (GB50176-93) [39], the city is classified within the cold climate zone of China (Figure 3). Xi'an experiences four distinct seasons: a mild spring, a hot summer, a cool autumn, and a cold winter [51].

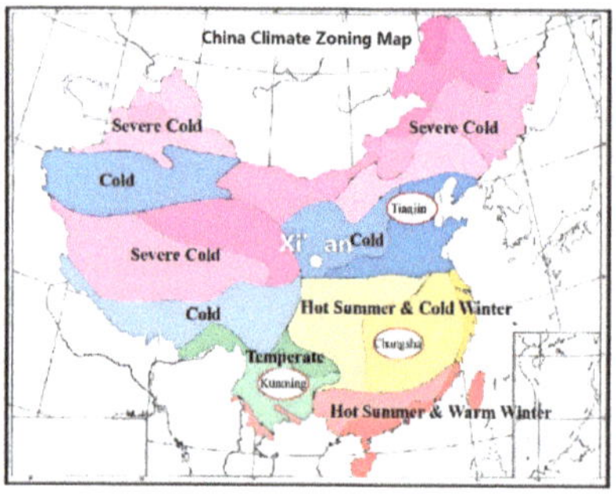

Figure 3. Detailed climate classification of Xi'an.

Meteorological data from 1971 to 2011 indicate that July has the highest average monthly temperature at approximately 27 °C, with a maximum recorded temperature of 31.7 °C. Conversely, January exhibits the lowest mean monthly temperature at −0.4 °C, with minimum temperatures reaching −3.4 °C. The annual mean relative humidity ranges between 57.2% and 77.4% [52].

2.2. The Analyzed Site

The field measurements were conducted on the Weishui Campus of Chang'an University, located in Xi'an, China. The test site was positioned within a corridor beneath an elevated building, featuring approximate dimensions of 15 m in length, 8.4 m in width, and 3.5 m in height (Figure 4).

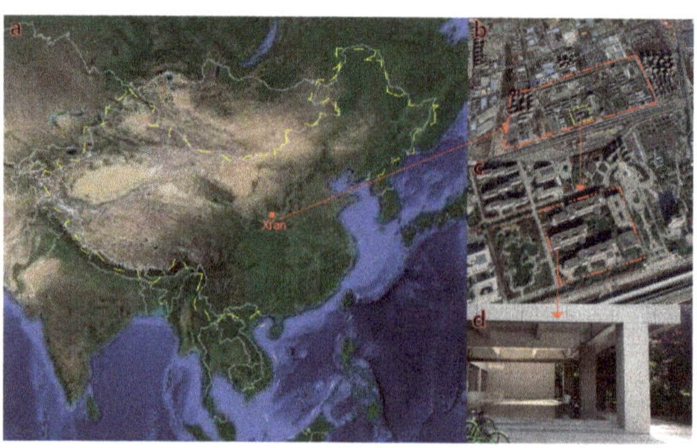

Figure 4. The research site in this study (**a**) Location of xi'an city (**b**) Weishui campus (**c**) Research area (**d**) Research site.

In this study, one measurement point was selected, located beneath the elevated structure (referred to as the elevated test point) (Figures 5 and 6). Meteorological parameters collected at this point were used for validating the ENVI-met simulation results. Boundary conditions for the ENVI-met simulations were derived from local meteorological data obtained from a nearby official weather station, ensuring representative and accurate input conditions for the model.

Figure 5. The selected points of this study.

Figure 6. The elevated point.

2.3. The Experimental Design

Field measurements were conducted separately during the summer and winter seasons, from 8 to 10 July 2020, and 4 to 6 January 2021, respectively, between 9:00 a.m. and 6:00 p.m. each day. These periods represent the typical hottest and coldest days of the year in Xi'an [53] (Table 1).

Table 1. The meteorological conditions during the measured days.

Measured Date	Mini Temperature	Max Temperature
8 July 2020	26 °C	36 °C
9 July 2020	25 °C	37 °C
10 July 2020	22 °C	29 °C
4 January 2021	−2 °C	3 °C
5 January 2021	−8 °C	1 °C
6 January 2021	−9 °C	0 °C

The monitored microclimatic parameters included air temperature (Ta), relative humidity (Rh), wind velocity (Va), and globe temperature (Tg). Air temperature and relative humidity were recorded using HOBO data loggers (accuracy: ±0.2 °C for air temperature and ±2.5% RH for relative humidity; resolution: 0.02 °C and 0.05% RH, respectively). Wind velocity was measured using the Kestrel 5500 anemometer [54], a fan-type device with an accuracy of ±3% of the reading or ±0.1 m/s, a measurement range of 0.6–40 m/s, and a resolution of 0.1 m/s. This anemometer provides reliable readings primarily above 0.6 m/s and may exhibit reduced accuracy at lower wind speeds typically encountered in outdoor thermal comfort assessments. Globe temperature was measured using the HD32.2 WBGT Index Instrument [54] (accuracy: ±0.1 °C; range: −10 to 100 °C; resolution: 0.1 °C) (Table 2).

Table 2. The detailed introduction of the instruments in this study.

Instrument	Parameter	Accuracy	Range	Resolution
Kestrel	Wind velocity	±3% of reading or ±0.1 m/s	0.6–40 m/s	0.1 m/s
HOBO	Air temperature Relative humidity	±0.2 °C ±2.5%	−40 °C ± 70 °C 0–100%	0.02 °C 0.05%
HD32.2 WBGT Index	Globe temperature	±0.1 °C	−10–100 °C	0.1 °C

We explicitly selected the 150 mm globe thermometer due to its higher measurement accuracy in quantifying radiant heat exchange, as extensively documented in the literature [54]. Although the 150 mm globe exhibits a comparatively longer response time than the 50 mm version, its improved accuracy in outdoor thermal comfort assessments justified this choice. All instruments were mounted at a standard pedestrian height and securely fixed to the ground. Measurements were recorded at one-minute intervals throughout the monitoring periods to capture detailed temporal variations.

2.4. The Introduction of the ENVI-Met Modeling

ENVI-met is a three-dimensional, non-hydrostatic microclimate simulation platform based on the fundamental principles of thermodynamics and fluid dynamics. It is specifically designed to evaluate the complex interactions among soil, vegetation, air, and buildings within urban environments. Figure 7 presents a detailed overview of the ENVI-met model layout used in this study. The simulation framework comprises several sub-models, including a one-dimensional (1D) boundary layer model, a three-dimensional (3D) soil model, and a 3D atmospheric model. The 1D model simulates vertical atmospheric processes from the ground surface (0 m) up to 2500 m in height. The 3D domain is discretized into grid cells, with spatial resolution defined in the model input files. To enhance computational accuracy near the pedestrian level, the lowest five vertical layers are refined using a stretching method, where the vertical increment is set to Δz = 0.2Δz.

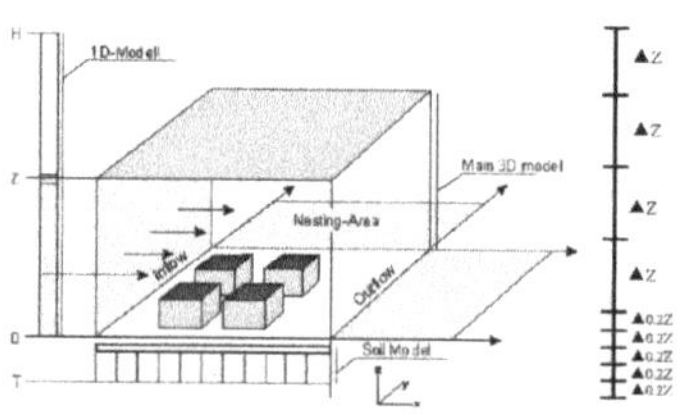

Figure 7. The overview of the ENVI-met.

In terms of resolution, ENVI-met supports a spatial grid size ranging from 0.5 to 10 m, and a temporal resolution from 1 to 10 s, allowing for flexible and high-fidelity simulations of urban microclimates [55].

Previous studies have extensively utilized ENVI-met to assess outdoor thermal environments and human thermal comfort in urban contexts [2,16,21,54–58]. A significant advancement in this regard is the release of a critical version of ENVI-met, which introduced the capability to import 3D models from Rhinoceros. This update addresses the limitations of earlier versions that lacked the flexibility to model complex urban forms, thereby making the tool more accessible and practical for architects and urban designers.

In the context of urban design and planning, it is essential to consider seasonal variations in solar radiation—particularly the need to block excessive solar gain during summer while permitting beneficial radiation during winter. This study leverages the latest ENVI-met features to simulate and analyze the thermal environment in semi-open spaces beneath elevated buildings during both summer and winter seasons.

A key climatic variable in this context is the solar altitude angle, which differs significantly between seasons. To take advantage of this, designers commonly incorporate architectural elements such as overhangs, awnings, and shutters, which serve to mitigate overheating in summer and enhance solar heat gain in winter.

Similarly, when designing elevated buildings, the thermal performance of the open space beneath the structure must also account for seasonal thermal variability. Identifying an optimal height range for these semi-outdoor spaces can substantially improve outdoor thermal comfort and contribute to reducing building energy consumption year-round (Figure 8).

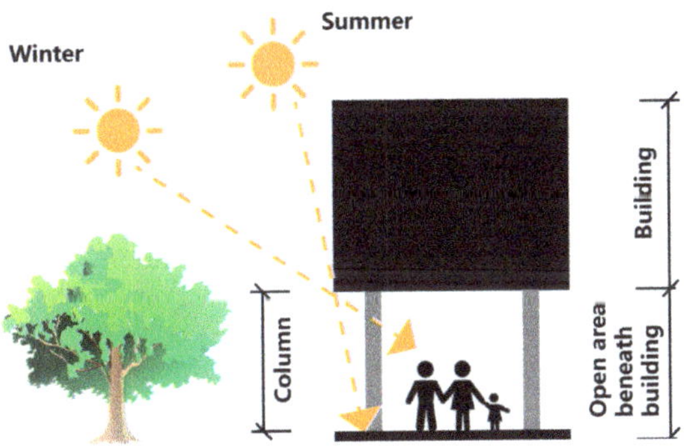

Figure 8. The design of the open area beneath the building.

2.5. The Framework of This Study

Figure 9 illustrates the overall research framework, which consists of three main steps:

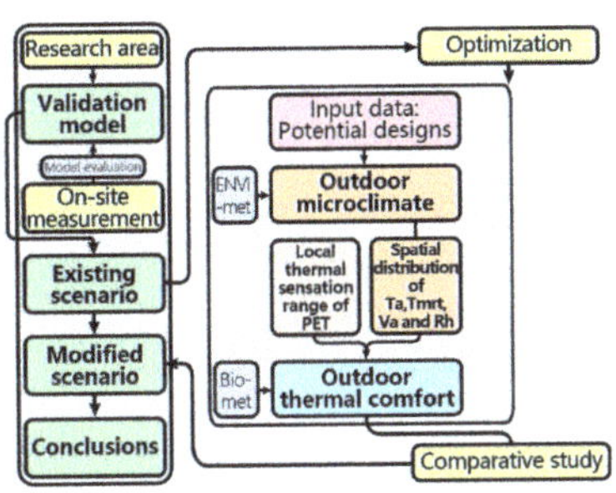

Figure 9. The framework of this study.

1. Field Measurements: On-site microclimatic data were collected during two distinct periods—8–10 July 2020 (summer) and 4–6 January 2021 (winter)—to capture seasonal variations in thermal conditions.
2. Model Validation: The ENVI-met simulation results were validated against the field measurements to ensure the reliability and accuracy of the model outputs.
3. Analysis and Conclusions: A detailed analysis of the simulation results was conducted to derive key findings, identify optimal design parameters, and formulate conclusions relevant to thermal comfort optimization in elevated open spaces.

3. Analyzed Results

3.1. The ENVI-Met Boundary Conditions

Field measurements collected over six consecutive days—covering both summer and winter seasons—were used to drive the ENVI-met simulations via the Simple Forcing option. The initial and boundary conditions applied in the simulations are detailed in Tables 3–9. These parameters were derived from the on-site field data and supplemented by meteorological records obtained from local weather stations [52].

Table 3. Initial settings of this study.

Parameters		Value	
Material properties (elevated surface)	Tile	Roughness length	0.01
		Albedo	0.5
		Emissivity	0.9
Boundary conditions	Air temperature Relative humidity	Tables 4–9	
	Turbulent model	Kinetic energy (TKE) model	
	Case-1 (Summer)		
	Wind speed(m/s)	1.5 m/s	
	Wind direction (°)	145	
	Grid in dx (m)	1	
	Grid in dy (m)	1	
	Grid in dz (m)	0.5	
	Number of x grid	200	
	Number of y grid	200	
	Number of z grid	30	
Simulation	Starting day	8–10 July 2020	
	Starting time	0:00 a.m.	
	Total simulation time	72 h	
	8 July	24 h	
	9 July	24 h	
	10 July	24 h	
Boundary conditions	Case-2 (Winter)		
	Wind speed (m/s)	1.3 m/s	
	Wind direction (°)	45	
	Grid in dx (m)	1	
	Grid in dy (m)	1	
	Grid in dz (m)	0.5	
	Number of x grid	200	
	Number of y grid	200	
	Number of z grid	30	

Table 3. *Cont.*

Parameters		Value
Simulation	Starting day	4–6 January 2021
	Starting time	0:00 a.m.
	Total simulation time	72 h
	4 January 5 January 6 January	24 h 24 h 24 h

Table 4. Initial air temperature and relative humidity in simulation (8 July).

Time	Air Temperature (°C)	Relative Humidity (%)	Time	Air Temperature (°C)	Relative Humidity (%)
0:00 a.m.	24.6	67.3	12:00 a.m.	35.4	38.8
1:00 a.m.	24	71	1:00 p.m.	36	37.5
2:00 a.m.	24.5	68.5	2:00 p.m.	36.3	36.1
3:00 a.m.	25.6	63.9	3:00 p.m.	36.9	34.5
4:00 a.m.	26.8	62	4:00 p.m.	36.1	34.9
5:00 a.m.	28.9	59.8	5:00 p.m.	35.2	35.9
6:00 a.m.	29.1	58.7	6:00 p.m.	34.2	38.8
7:00 a.m.	30.2	58.1	7:00 p.m.	32.5	45.6
8:00 a.m.	31	54.8	8:00 p.m.	29.8	48.3
9:00 a.m.	31.6	54.5	9:00 p.m.	27	52.8
10:00 a.m.	33.4	44.9	10:00 p.m.	25.9	61.8
11:00 a.m.	34.6	42.3	11:00 p.m.	25.1	62.5

Table 5. Initial air temperature and relative humidity in simulation (9 July).

Time	Air Temperature (°C)	Relative Humidity (%)	Time	Air Temperature (°C)	Relative Humidity (%)
0:00 a.m.	27	65.1	12:00 a.m.	34.8	39.8
1:00 a.m.	26.3	71.1	1:00 p.m.	35.8	37.6
2:00 a.m.	26.8	68.3	2:00 p.m.	36.5	36
3:00 a.m.	27.6	66.5	3:00 p.m.	36.9	35.5
4:00 a.m.	27.9	62.1	4:00 p.m.	36.6	36.2
5:00 a.m.	28.3	59.5	5:00 p.m.	36.1	38.1
6:00 a.m.	28.9	58.1	6:00 p.m.	35.3	39.1
7:00 a.m.	29.4	56.3	7:00 p.m.	34.1	40.5
8:00 a.m.	29.9	55.1	8:00 p.m.	32.3	42.1
9:00 a.m.	30.6	53.2	9:00 p.m.	30.8	48.6
10:00 a.m.	33.3	44.9	10:00 p.m.	28.9	55.1
11:00 a.m.	33.8	40.9	11:00 p.m.	27.8	61.2

Table 6. Initial air temperature and relative humidity in simulation (10 July).

Time	Air Temperature (°C)	Relative Humidity (%)	Time	Air Temperature (°C)	Relative Humidity (%)
0:00 a.m.	24.1	72.2	12:00 a.m.	29.9	61.2
1:00 a.m.	23.2	75.1	1:00 p.m.	30.1	58.9
2:00 a.m.	24.2	73.1	2:00 p.m.	31.2	58
3:00 a.m.	24.9	71.5	3:00 p.m.	32	57
4:00 a.m.	25.6	69.9	4:00 p.m.	31	57.9
5:00 a.m.	26.9	68.1	5:00 p.m.	30.5	58.9
6:00 a.m.	27.1	67	6:00 p.m.	30	59.3
7:00 a.m.	27.8	66.1	7:00 p.m.	28.5	60.2
8:00 a.m.	28.3	65.1	8:00 p.m.	27.1	63.1
9:00 a.m.	28.6	64.8	9:00 p.m.	26.5	65.6
10:00 a.m.	28.9	63.9	10:00 p.m.	25.4	67.8
11:00 a.m.	29	62.7	11:00 p.m.	24.8	69.1

Table 7. Initial air temperature and relative humidity in simulation (4 January).

Time	Air Temperature (°C)	Relative Humidity (%)	Time	Air Temperature (°C)	Relative Humidity (%)
0:00 a.m.	−2.1	51.2	12:00 a.m.	2.91	44.08
1:00 a.m.	−2.5	53.1	1:00 p.m.	3.70	42.96
2:00 a.m.	−1.9	50.8	2:00 p.m.	3.99	44.12
3:00 a.m.	−1.5	49.7	3:00 p.m.	4.40	44.43
4:00 a.m.	−1.1	49.6	4:00 p.m.	5.16	42.96
5:00 a.m.	−0.63	49.1	5:00 p.m.	4.32	44.31
6:00 a.m.	0.16	48.8	6:00 p.m.	3.27	47.41
7:00 a.m.	0.95	48.5	7:00 p.m.	2.51	48.1
8:00 a.m.	1.23	48.4	8:00 p.m.	1.68	48.93
9:00 a.m.	1.77	47.99	9:00 p.m.	0.91	49.21
10:00 a.m.	1.88	47.94	10:00 p.m.	−0.85	49.6
11:00 a.m.	2.32	44.83	11:00 p.m.	−1.63	49.9

Table 8. Initial air temperature and relative humidity in simulation (5 January).

Time	Air Temperature (°C)	Relative Humidity (%)	Time	Air Temperature (°C)	Relative Humidity (%)
0:00 a.m.	−3.9	39.5	12:00 a.m.	5.52	26.92
1:00 a.m.	−4.5	39.9	1:00 p.m.	5.52	26.47
2:00 a.m.	−3.5	39.1	2:00 p.m.	6.41	26.36
3:00 a.m.	−2.6	38.8	3:00 p.m.	6.40	26.89
4:00 a.m.	−1.7	38.5	4:00 p.m.	6.43	26.49
5:00 a.m.	−0.86	38	5:00 p.m.	6.87	29.36
6:00 a.m.	0.35	36.1	6:00 p.m.	5.31	31.36
7:00 a.m.	1.15	33.5	7:00 p.m.	3.85	33.2
8:00 a.m.	2.35	31.5	8:00 p.m.	2.56	34.8
9:00 a.m.	3.47	30.17	9:00 p.m.	1.68	36.12
10:00 a.m.	3.32	29.76	10:00 p.m.	0.98	36.85
11:00 a.m.	4.53	27.88	11:00 p.m.	−1.12	38.5

Table 9. Initial air temperature and relative humidity in simulation (6 January).

Time	Air Temperature (°C)	Relative Humidity (%)	Time	Air Temperature (°C)	Relative Humidity (%)
0:00 a.m.	−8.3	32.1	12:00 a.m.	0.91	24.1
1:00 a.m.	−9.3	33.5	1:00 p.m.	0.95	23.9
2:00 a.m.	−7.1	30.4	2:00 p.m.	1.10	23.1
3:00 a.m.	−5.1	29.7	3:00 p.m.	1.48	22.8
4:00 a.m.	−4.2	29	4:00 p.m.	1.25	21.5
5:00 a.m.	−3.5	28.15	5:00 p.m.	0.69	21.4
6:00 a.m.	−2.4	25.3	6:00 p.m.	0.30	21.63
7:00 a.m.	−0.8	25.1	7:00 p.m.	−1.36	26.5
8:00 a.m.	0.1	24.9	8:00 p.m.	−3.51	27.85
9:00 a.m.	0.80	24.34	9:00 p.m.	−4.15	28.91
10:00 a.m.	0.85	25.40	10:00 p.m.	−5.89	29.9
11:00 a.m.	0.80	22.79	11:00 p.m.	−7.1	30.5

3.2. The Validation Between the Measured and Simulated Data

The accuracy of the simulation was assessed using a set of statistical metrics commonly applied in outdoor thermal environment studies. Specifically, we evaluated the model performance using Root Mean Square Error (RMSE), Systematic Root Mean Square Error (RMSEs), Unsystematic Root Mean Square Error (RMSEu), and the coefficient of determination (R^2). These metrics were calculated at a representative point located beneath the elevated building. To ensure the reliability of the simulation results, the following criteria are generally recommended: RMSE and RMSEs should approach 0, RMSEu should approximate RMSE, and R^2 should be close to 1 [59]. These indicators provide a robust basis for evaluating the model's predictive capability. As discussed earlier, thermal indices have become widely adopted for predicting outdoor human thermal comfort. Among them, Mean Radiant Temperature (MRT)—also denoted as TMRT—plays a crucial role. MRT is defined as the uniform temperature of an imaginary enclosure in which the radiant heat exchange between a human body and the surrounding surfaces is equivalent to that in a real, non-uniform environment. The calculation of MRT follows the methodology specified in ISO 7726 [60].

$$T_{MRT} = \left[(Tg + 273)^4 + \frac{1.10 \times Va \times 10^8}{\varepsilon \times \mathrm{D}^{0.4}}(\mathrm{Tg} - \mathrm{Ta})\right]^{\frac{1}{4}} - 273 \quad (1)$$

where Tg is the globe temperature (°C), Ta is the air temperature (°C), Va is the wind speed (m/s) at the selected point, D is the globe diameter (0.05 m in this study), and ε is the emissivity of the globe (0.95 for a black globe).

The PET index in ENVI-met is calculated based on several key variables, including Ta, Rh, Va, and TMRT. Therefore, validating the accuracy of these simulated parameters is a critical step in ensuring the reliability of the PET output. Figure 10 and Table 10 present the comparison between measured and simulated data for the summer period. The simulation of Ta demonstrated high precision across all measured days, with an average R^2 of 0.9008 and a RMSE of 1.22 °C. Similarly, Rh and TMRT showed strong agreement with measured values, with R^2 values of 0.8573 and 0.8653, and RMSE values of 6.86% and 3.12 °C, respectively—indicating consistent model performance.

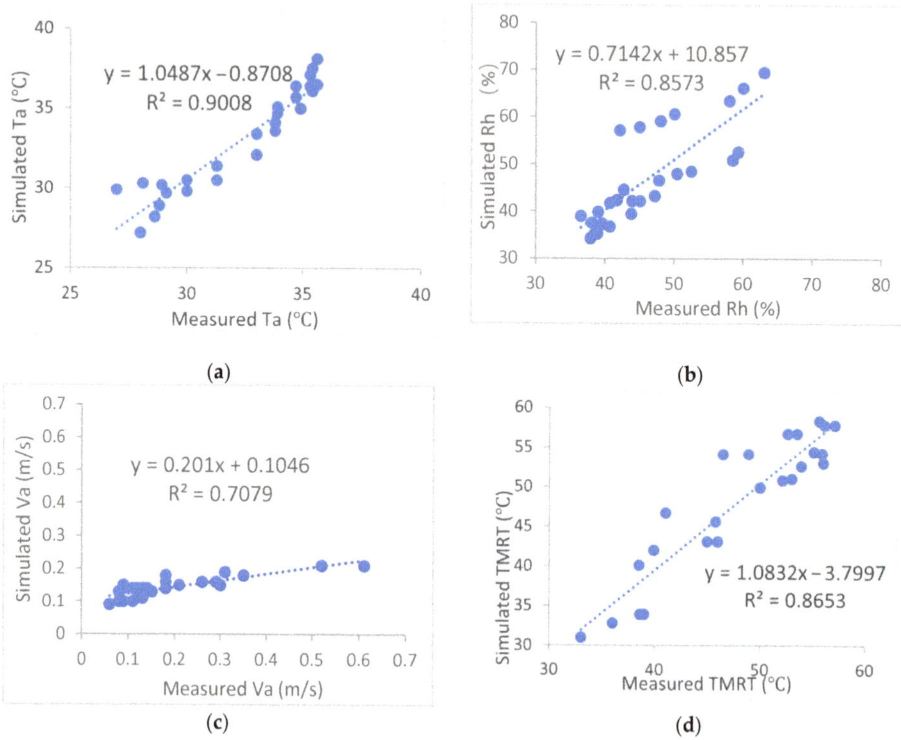

Figure 10. The R^2 of the measured and simulated data of the elevated point in summer (**a**) Ta, (**b**) Rh, (**c**) Va, and (**d**) TMRT.

Table 10. The quantitative evaluation of the ENVI-met performance on the evaluation of summer in RMSE, RMSEu, and RMSEs.

Parameter	RMSE	RMSEu	RMSEs
Ta	1.22 °C	1.98 °C	1.03 °C
Rh	6.86%	7.02%	4.85%
Va	0.11 m/s	0.08 m/s	0.065 m/s
TMRT	3.12 °C	3.26 °C	1.89 °C

Although the simulation accuracy for Va was relatively lower than for other parameters, the model still provided a reliable estimation of wind conditions, consistent with previous validation results [16].

For the simulated microclimate in winter (Figure 11, Table 11), the validation results show a similarly high level of accuracy as observed in the summer. The R^2 values for Ta, Rh, Va, and TMRT were 0.814, 0.819, 0.6754, and 0.9633, respectively. Correspondingly, the RMSE values were 1.22 °C, 6.86%, 0.11 m/s, and 3.12 °C, respectively. These results confirm that the simulation model performs reliably in winter conditions, particularly for Ta and TMRT, while Va exhibits slightly lower predictive accuracy.

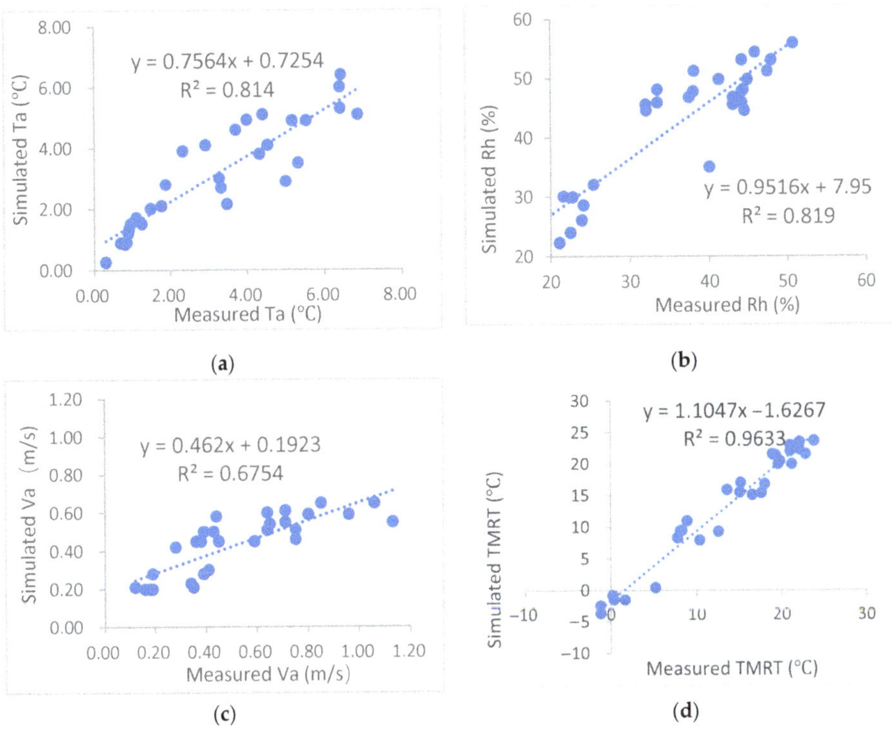

Figure 11. The R^2 of the measured and simulated data of the elevated point in winter (**a**) Ta, (**b**) Rh, (**c**) Va, and (**d**) TMRT.

Table 11. The quantitative evaluation of the ENVI-met performance on the evaluation of winter in RMSE, RMSEu, and RMSEs.

Parameter	RMSE	RMSEu	RMSEs
Ta	0.88 °C	1.21 °C	0.69 °C
Rh	7.65%	8.3%	6.59%
Va	0.19 m/s	0.13 m/s	0.11 m/s
TMRT	1.91 °C	2.09 °C	1.96 °C

These findings confirm the model's applicability in supporting climate-responsive design strategies for elevated structures in cold climate urban environments.

3.3. The PET Values of the Existing Scenario

As previously discussed, the thermal comfort index PET has become widely adopted for evaluating outdoor thermal environments. A prior study [61] (Table 12) established a PET-based thermal perception scale specifically tailored for the cold climate zone of northern China. This 9-point classification system includes the categories: Very Cold, Cold, Cool, Slightly Cool, Neutral, Slightly Warm, Warm, Hot, and Very Hot, each corresponding to a specific PET range. This scale is employed in the present study to assess outdoor thermal comfort conditions beneath elevated buildings.

Table 12. The PET values in the cold climate zone of northern China [61].

Thermal Stress	Thermal Sensation	PET (°C)
Extreme cold stress	Very Cold	<−4
Strong cold stress	Cold	−4~8
Moderate cold stress	Cool	8~16
Slightly cold stress	Slightly Cool	16~22
No thermal stress	Neutral	22~28
Slight heat stress	Slightly Warm	28~32
Moderate heat stress	Warm	32~38
Strong heat stress	Hot	38~44
Extreme heat stress	Very Hot	>44

In ENVI-met, the default summer settings define sunrise at 6:00 a.m. and sunset at 7:00 p.m. The simulated PET distribution for the summer scenario is presented in Figure 12. As shown, all three curves exhibit similar trends, with PET values increasing after sunrise and peaking around 2:00 p.m., followed by a gradual decline. During the core observation period from 9:00 a.m. to 6:00 p.m., the site consistently falls within the 'Hot' to 'Very Hot' PET categories, indicating significant thermal stress throughout most of the daytime period.

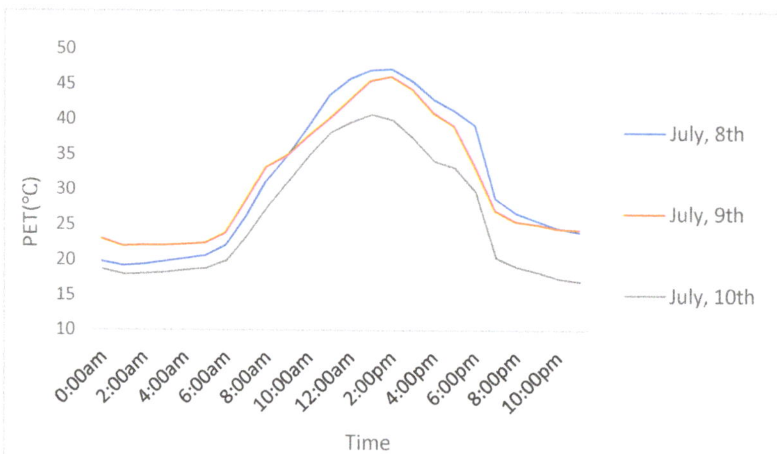

Figure 12. The PET values of the researched site in summer.

Figure 13 illustrates the variation of PET during the winter season. Throughout the daytime period, the entire site predominantly remains within the 'Cool' category, indicating moderate thermal discomfort. In contrast, during nighttime hours, PET values drop significantly, falling into the 'Cold' and 'Very Cold' categories, reflecting the harsh thermal conditions characteristic of winter nights in cold climate zones.

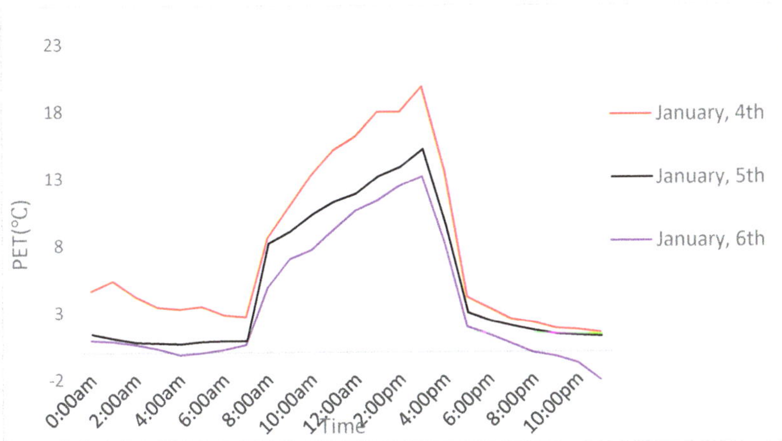

Figure 13. The PET values of the researched site in winter.

4. Discussion

As summarized in the preceding sections, further discussion is warranted regarding the thermal comfort characteristics of the semi-open space beneath elevated buildings and the corresponding design implications. A key structural component in such configurations is the supporting column, which plays a critical role not only in the load-bearing function but also in shaping the local microclimate. Therefore, the discussion begins by evaluating the influence of column dimensions and placement on the surrounding thermal environment.

4.1. The Impact of the Columns on the Surrounding Thermal Environment

Due to the default grid constraints in ENVI-met, the minimum allowable length and width for model elements is 0.5 m, and all dimensions must be in multiples of this unit. Based on this limitation, a series of simulation cases was constructed using columns with varying widths to specifically evaluate their influence on the surrounding thermal environment, independently of seasonal conditions. In total, five scenarios were modeled, incorporating column widths of 0.5 m, 1.0 m, 1.5 m, 2.0 m, and 2.5 m. All other parameters and boundary conditions were held constant, explicitly to isolate the effect of column dimension alone, rather than assess seasonal differences or variations. The corresponding simulation results, solely focused on verifying the influence of column size, are presented in Table 13 and Figure 14.

The results indicate that PET values increase slightly as column width increases; however, the differences are marginal. Specifically, when the column width increases from 0.5 m to 2.5 m, the maximum PET difference observed is only 0.6 °C. This minimal variation clearly suggests that column width itself has a negligible impact on the thermal comfort of the surrounding environment. Therefore, within the design constraints of typical elevated buildings, column dimension can be considered a secondary factor in thermal performance optimization.

Table 13. Simulated PET results under varying column widths.

Column Width	PET
0.5 m	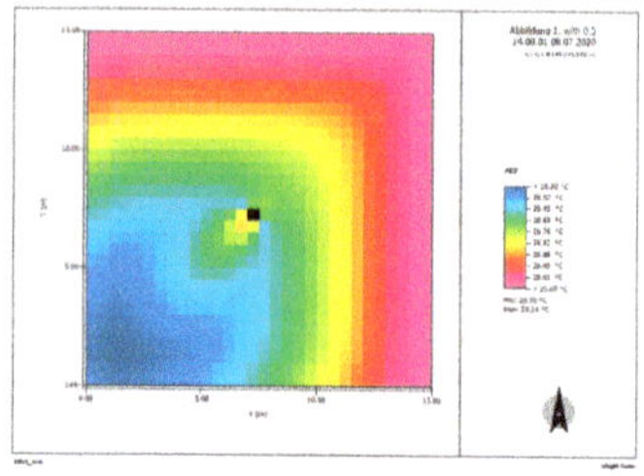
1 m	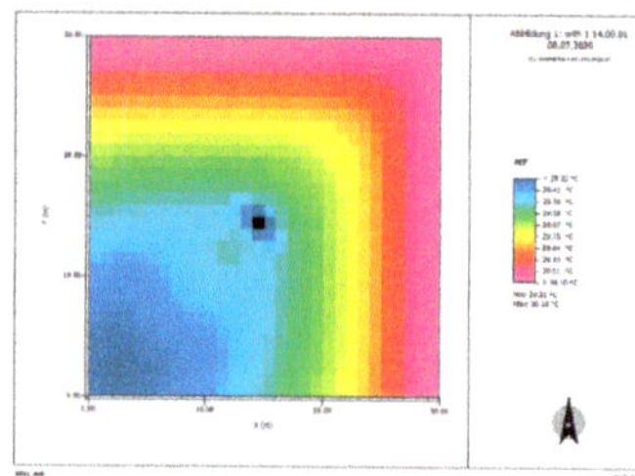
1.5 m	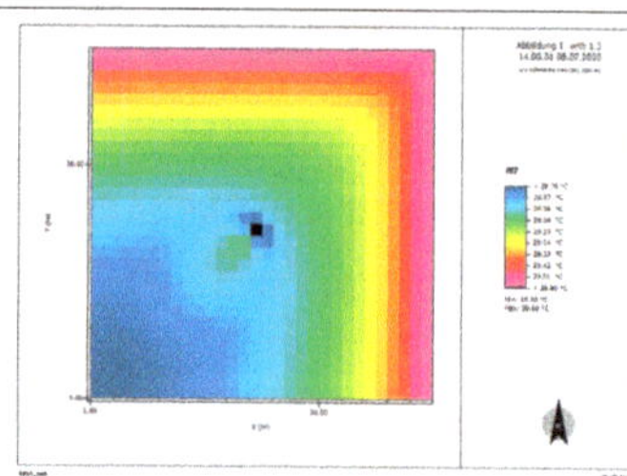
2 m	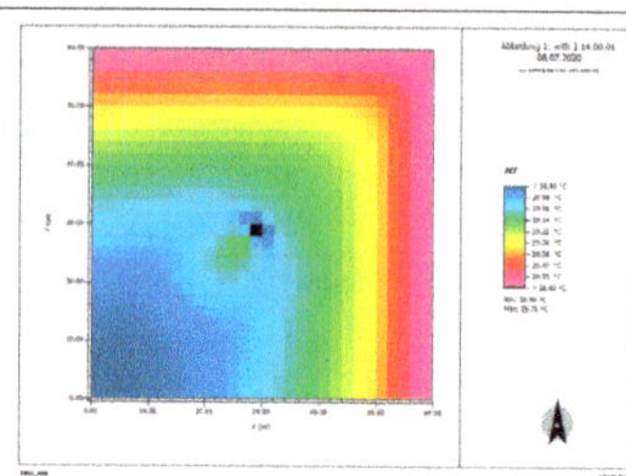
2.5 m	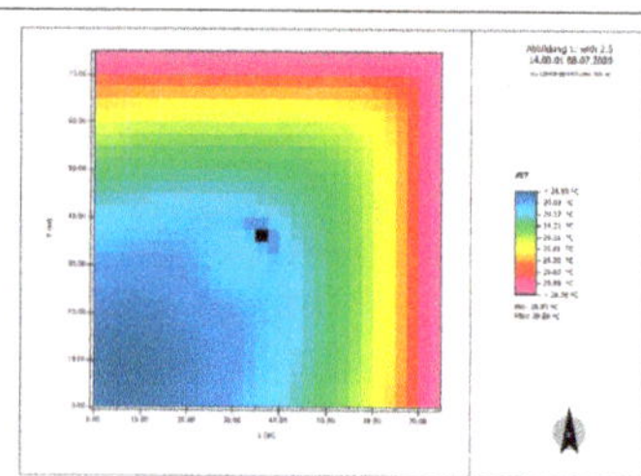

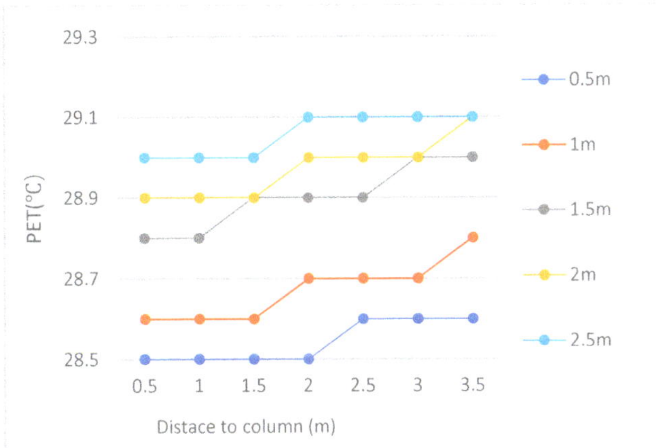

Figure 14. PET values simulated under varying column widths.

4.2. The Impact of the Bottom of the Elevated Building on the Surrounding Thermal Environment

Following the analysis of column width, it is also necessary to explore the broader impact of column-supported elevated structures on the thermal environment beneath buildings. To this end, two hypothetical cases were developed based on the same base geometry (Table 14).

Table 14. The different hypothesis cases.

Base Case	Case-1	Case-2
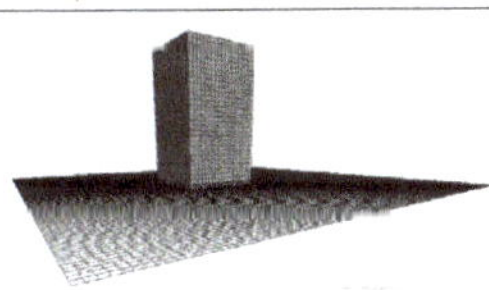	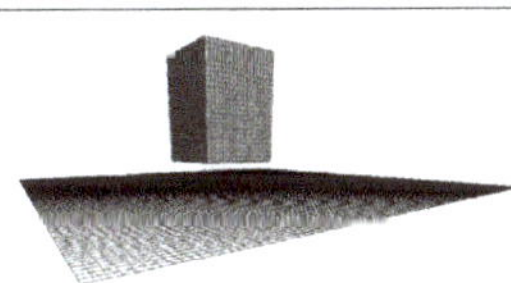	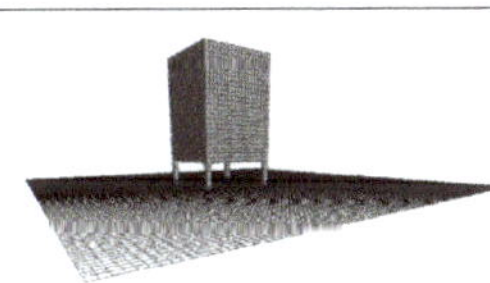

Case 1 represents a theoretical scenario in which the first floor is completely removed, resulting in a fully open ground surface without any overhead structure. Although this configuration does not occur in real-world architecture, it serves as a comparative reference to isolate the effect of the elevated structure itself. Case 2 simulates a typical elevated ground floor supported by columns—reflecting a standard design commonly used in high-density urban developments.

The simulated PET results for both cases are presented in Table 15. The comparison provides insights into how the presence of elevated structures and their supporting columns influences microclimatic conditions and thermal comfort at the pedestrian level.

The simulation results for Case 1 and Case 2 reveal highly similar thermal conditions in the selected area. Specifically, Ta and PET values in Case 1—where the elevated floor is entirely absent—are nearly identical to those in Case 2, which features a typical column-supported elevated design. This finding suggests that the presence of supporting columns has minimal influence on the local thermal environment and outdoor thermal comfort beneath elevated buildings. The comparative results are illustrated in Figure 15.

Table 15. The simulated result of different hypothesis cases.

Case	PET
Case-1	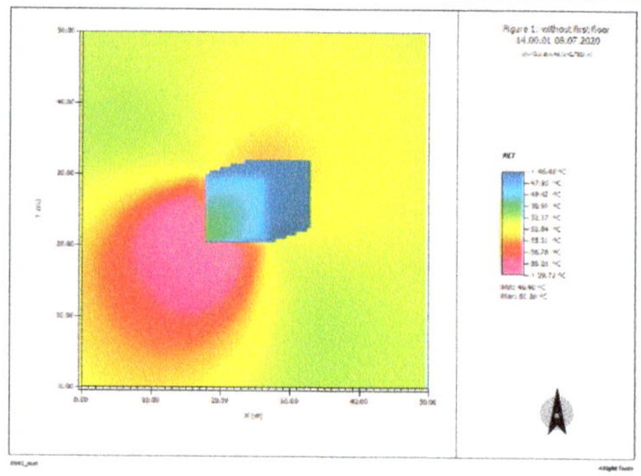
Case-2	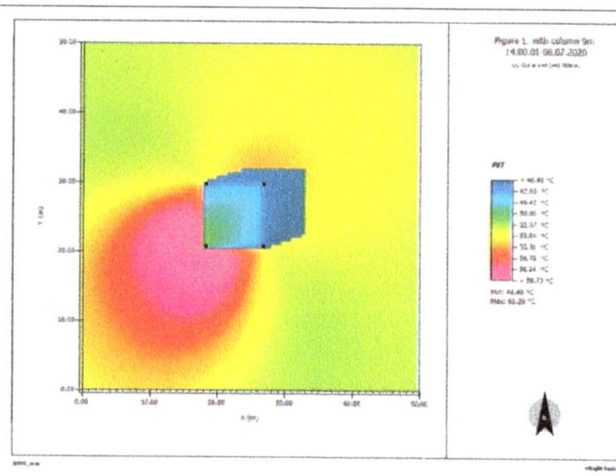

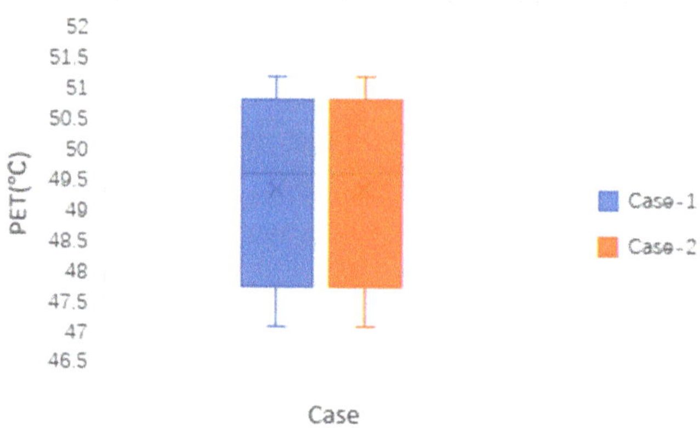

Figure 15. The Ta and PET values in two different cases.

4.3. *The Difference Between the Overall Elevated Design and the Partial Elevated Design*

In addition, the elevation ratio—defined as the proportion of the ground floor area that is elevated—plays an important role in influencing the thermal environment and outdoor comfort. In this section, four scenarios were modeled with elevation ratios of 20%, 50%, 80%, and 100%, respectively, to examine their impact on microclimatic conditions. Table 16 and Figure 16 present the distributions of Ta and PET across the four cases. The comparison helps to reveal how varying degrees of elevation affect thermal performance beneath the building structure.

The simulation results clearly indicate that increasing the elevation ratio leads to an amplification of thermally comfortable areas, primarily due to the enhanced shading effect provided by the elevated structure. As shown in Figure 16, the minimum, average, and maximum values of Ta and PET exhibit similar trends across the four scenarios.

Among the tested cases, the 40% and 70% elevation ratios result in slightly higher Ta and PET values compared to the 10% and 100% cases. This can be attributed to reflected radiation from surrounding building surfaces, which partially offsets the shading benefit. Nevertheless, the 100% elevation case provides the most consistently comfortable thermal environment overall, as the continuous shading significantly reduces solar exposure at the pedestrian level.

Table 16. The simulated result of different hypothesis cases.

Case	PET
10%-case	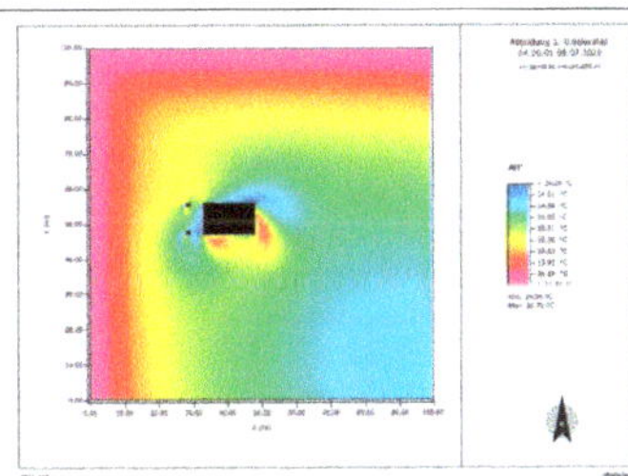
40%-case	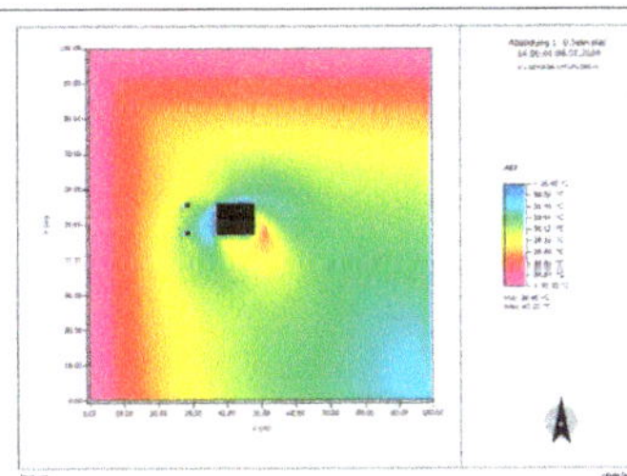
70%-case	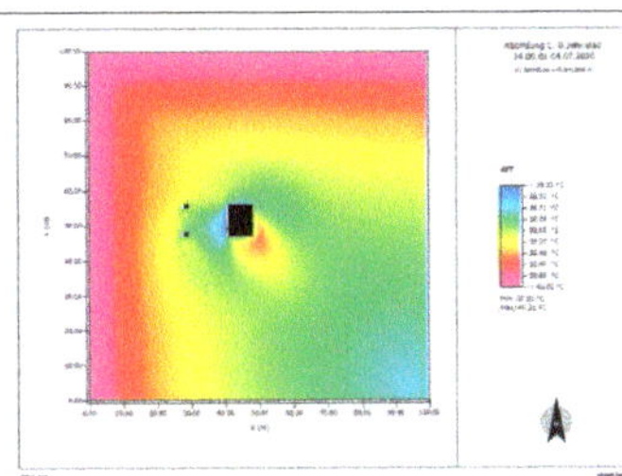
100%-case	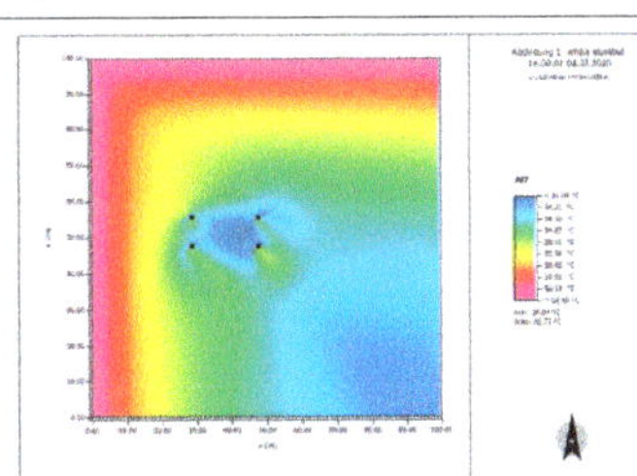

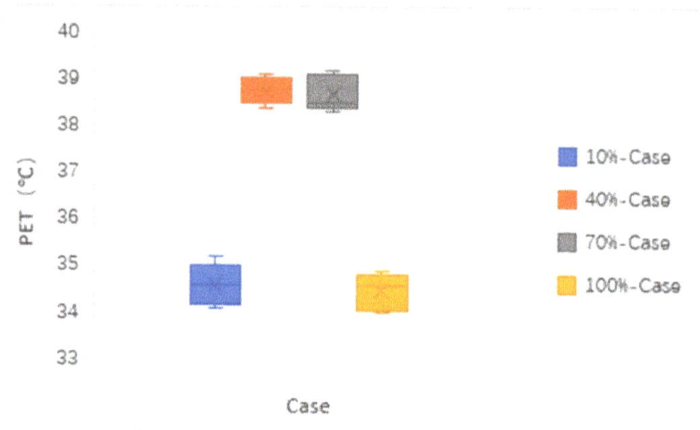

Figure 16. The Ta and PET values in four different cases.

To further clarify the relationship between the elevation ratio and the surrounding thermal environment, a numerical correlation was established between the elevation ratio and the average PET, as illustrated in Figure 17. The analysis reveals that the relationship follows a second-degree polynomial trend, rather than a simple linear correlation. The fitted equation for PET is expressed as Equation (2):

$$PET = -22.5x^2 + 24.85x + 22.295 \quad (2)$$

$$R^2 = 0.9963$$

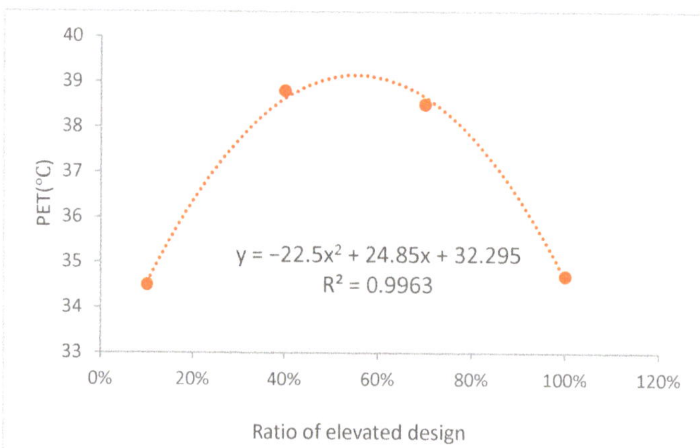

Figure 17. The relationship between the Elevated ratio and average PET.

This result suggests that thermal comfort initially deteriorates as the elevation ratio increases. However, when the ratio exceeds approximately 55%, PET values begin to decrease, indicating an improvement in thermal comfort due to increased and more continuous shading effects.

4.4. The Estimation of the Optimal Height of the Elevated Building

Given that China is located in the Northern Hemisphere, local architectural design standards emphasize maximizing natural daylighting within buildings. As a result, east–

west building orientations are generally preferred. In addition, to accommodate pedestrian circulation and recreational activities, the minimum required height for open spaces beneath elevated buildings is set at no less than 2 m, and the ideal length-to-width ratio is recommended to be 2:1 [62].

Based on these regulatory considerations, combined with the minimum spatial resolution constraint of 0.5 m in ENVI-met modeling, a total of eight simulation scenarios were developed. These scenarios feature elevated space heights of 2.0 m, 2.5 m, 3.0 m, 3.5 m, 4.0 m, 4.5 m, 5.0 m, 5.5 m, and 6.0 m, respectively. The plan dimensions of the teaching building were limited to a maximum length of 50 m and a width of 10 m to align with standard design practice. For each scenario, meteorological data from the hottest day (9 July 2020) and the coldest day (6 January 2021) were used as initial boundary conditions for ENVI-met simulations.

To effectively compare the thermal environment across all height configurations, box plots were used to visualize the distribution of PET values throughout the elevated space. The summer simulation results reveal a negative correlation between the elevation height and thermal comfort: as the elevation height increases, PET values rise, indicating a decline in thermal performance. For instance, the 6.0 m case recorded a peak PET of 40.5 °C, while the 2.0 m case showed a maximum PET of 37.5 °C. According to the thermal comfort classification in Table 12, this range corresponds to a shift in perceived thermal conditions from the 'Warm' to the 'Hot' category (Figure 18).

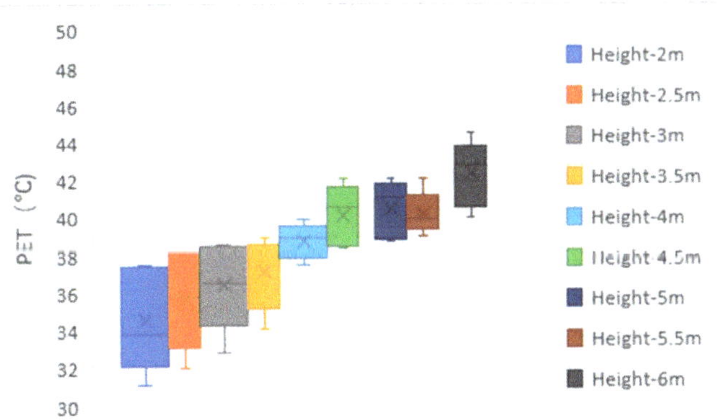

Figure 18. PET Distribution Across Different Elevated Heights During Summer.

In contrast, winter simulation results exhibit a similar pattern of PET change with height but differ in comfort implications. As shown in Figure 19, the 2.0 m case provides the least comfortable thermal environment in winter, while the 6.0 m case yields the highest thermal comfort. Based on the comfort classification presented in Table 12, all cases generally fall within the 'Cold' category, but the taller elevated spaces improve solar penetration and air circulation, enhancing thermal perception.

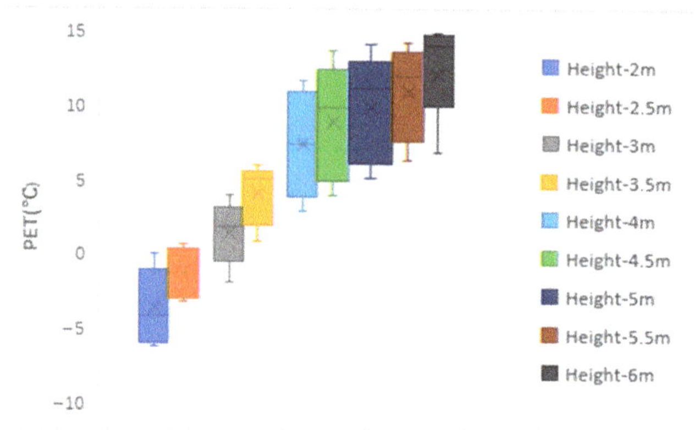

Figure 19. PET Distribution Across Different Elevated Heights During Winter.

Earlier studies have shown that elevated building designs can enhance thermal comfort during summer by increasing airflow and providing effective shading. As previously stated, the present study focuses on further investigating the thermal performance of the entire space beneath elevated buildings, offering data-driven insights to support climate-responsive architectural design.

Based on the PET classification in Table 12, thermal comfort levels are grouped into three categories:

- Unsuitable: Very Cold, Cold, Very Hot, Hot
- Fairly Suitable: Cool, Warm
- Suitable: Slightly Cool, Neutral, Slightly Warm

Given the inherent heat in summer, achieving the Neutral comfort stage is unrealistic. Therefore, this study designates the Slightly Warm and Warm categories as acceptable levels of outdoor thermal comfort for the summer season.

To identify the optimal elevated height for summer conditions, we applied regression analysis to the simulation data. As shown in Figure 20, the relationship between elevated height and average PET in summer follows a linear trend, expressed as:

$$PET = 2.0033x + 29.387 \tag{3}$$

$$R^2 = 0.9884$$

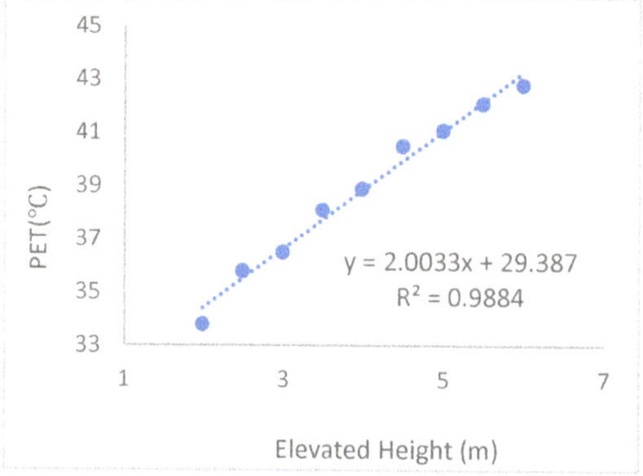

Figure 20. Relationship Between Elevated Height and Average PET in Summer Conditions.

With a coefficient of determination $R^2 = 0.9884$, indicating a highly significant correlation. Based on this equation and the PET thresholds defined in Table 12, the ideal elevation height for achieving acceptable summer thermal comfort falls between −0.73 m and 4.3 m. However, considering practical construction constraints and regulatory guidelines [62], the feasible and recommended elevation height range is 2.0 m to 4.3 m.

A similar method was applied to evaluate thermal comfort during winter. For this season, Slightly Cool and Cool are considered acceptable comfort levels. The corresponding regression equation, depicted in Figure 21, is:

$$PET = 4.998x - 10.045 \quad (4)$$

$$R^2 = 0.87$$

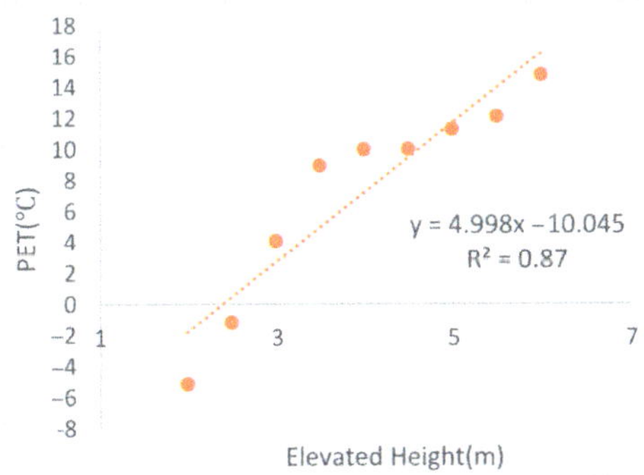

Figure 21. Relationship Between Elevated Height and Average PET in Winter.

The corresponding regression depicted in Figure 21 has a coefficient of determination $R^2 = 0.87$, also indicating a meaningful fit. According to this model, the optimal elevated height for winter lies between 3.6 m and 6.4 m.

By comparing the results from both summer and winter simulations, it becomes clear that achieving thermal balance in elevated open spaces requires careful calibration. In the cold climate zone of Northern China—where mitigating heat stress in summer and retaining warmth in winter are both essential—the recommended optimal year-round elevation height falls within the range of 3.6 m to 4.3 m.

It is valuable to contextualize the PET results obtained in this study by comparing them with similar investigations from other cold-climate regions in China. Recent research conducted in Harbin and Beijing has reported typical PET comfort thresholds ranging between 18–23 °C for summer and around 4–8 °C for winter [63]. These regional benchmarks align closely with the optimal ranges identified by our findings.

Furthermore, considering that Xi'an is geographically situated near the boundary between the "Cold" and "Hot summer–Cold winter" climate zones according to the Chinese national climate classification (GB 50176-2016) [39], this location inherently introduces a complexity that may affect the generalizability of the results. To address this complexity explicitly, we have further compared our outcomes with recent studies conducted in cities explicitly characterized by the "Hot summer–Cold winter" climate zone, such as Wuhan and Nanjing. These comparisons reveal that our recommended optimal elevated heights and spatial design thresholds are consistent with the findings from these mixed climate cities. This consistency is particularly evident regarding the architectural challenge of balancing shading effectiveness during hot summers with solar access in cold winters, reinforcing the robustness and practical relevance of our proposed design recommendations.

Moreover, previous studies specifically focusing on elevated or semi-open spaces beneath buildings in similar climatic conditions have demonstrated that these architectural configurations typically improve outdoor thermal comfort by approximately 5–10 °C compared to fully exposed outdoor spaces [64]. Such findings further substantiate and reinforce our conclusion that properly designed elevated building spaces can significantly enhance pedestrian comfort in regions experiencing distinct seasonal extremes.

Integrating these insights with national building design standards and modular dimension guidelines, our study refines the proposed optimal elevation thresholds to 3.6 m, 3.9 m, and 4.2 m. These recommended values constitute a balanced solution that simultaneously meets thermal efficiency requirements and practical feasibility constraints, thus providing valuable and directly applicable guidelines for the climate-adaptive design of elevated buildings in regions characterized by transitional or mixed climatic conditions.

5. Conclusions

This study systematically evaluated the microclimatic performance of elevated ground-floor semi-open spaces beneath buildings in the cold climate zone of Northern China, using field-validated ENVI-met simulations. Based on comprehensive quantitative analyses, the following clear conclusions and practical implications were derived:

(1) Structural dimensions and thermal comfort implications:

Our results indicated that variations in supporting column widths (ranging from 0.5 m to 2.5 m) had minimal impact on surrounding thermal conditions, with PET differences remaining within a negligible range (maximum of 0.6 °C). This finding implies that architects and structural engineers have substantial flexibility in selecting column dimensions, allowing decisions to prioritize structural requirements, aesthetic considerations, or functional needs over minor thermal comfort influences.

(2) Optimal elevated ratio for thermal comfort:

For buildings oriented in an east–west configuration—typical in Northern China for daylight optimization—the relationship between elevation ratio and thermal comfort (PET) was found to be nonlinear, following a clear parabolic trend. Specifically, PET initially increased (indicating reduced comfort) as elevation ratios increased, reaching a peak discomfort of around 55%. Beyond this critical threshold, further increases in elevation ratio significantly improved thermal comfort due to enhanced shading effectiveness. These findings offer valuable guidance for architectural and urban planning practice, recommending higher elevation ratios ($\geq$55%) to effectively enhance outdoor pedestrian comfort in urban environments.

(3) Recommended year-round elevated height:

Our analysis demonstrated that optimal elevation heights varied seasonally, with an ideal range of 2.0 m to 4.3 m for summer thermal comfort and 3.6 m to 6.4 m for winter comfort. Considering both seasonal extremes together, the recommended year-round elevation height range is 3.6 m to 4.3 m. Moreover, consistent with modular construction principles and national building design standards, specific elevation heights of 3.6 m, 3.9 m, and 4.2 m were identified as particularly advantageous, balancing thermal efficiency, construction practicality, and regulatory compliance. These clear and actionable design thresholds can be directly applied by practitioners aiming to optimize thermal comfort in architectural and urban projects.

Overall, this study contributes meaningful evidence-based recommendations for climate-responsive design practice. Although conducted specifically in Xi'an, the methodological framework and validated findings possess considerable potential for broader application across similar cold-climate urban settings. Future research is encouraged to expand these analyses into other urban locations, climate contexts, and additional seasons (spring and autumn), further enhancing the robustness, relevance, and global applicability of our findings.

6. Limitations and Outlook

This study has provided a detailed evaluation of the microclimatic performance of elevated semi-open spaces beneath buildings in Xi'an using traditional statistical methods and ENVI-met simulations. Despite the valuable insights gained, several limitations should be explicitly acknowledged. Firstly, the current analysis was conducted at a single urban location, limiting the generalizability of our conclusions. Secondly, the assessment focused exclusively on two seasonal extremes (summer and winter), excluding transitional seasons (spring and autumn), which could offer additional meaningful insights into thermal comfort dynamics.

Additionally, the methodological approach employed in this study primarily relies on conventional statistical methods, which, although robust and widely accepted, have yet to fully incorporate the advantages of emerging technologies. With rapid developments in artificial intelligence (AI) and advanced computational frameworks, there are significant opportunities to enhance the precision, scalability, and automation capabilities of environmental modeling in urban design research.

In future research, we propose expanding our analytical scope to multiple urban locations and comprehensive seasonal assessments across diverse climatic zones, thus improving the robustness and broad applicability of the findings. Moreover, we plan to integrate AI-driven approaches by establishing cross-platform linkages between Python 3.12.6 and ENVI-met 5.0. This innovative integration will allow advanced computational analyses, automated optimization, and adaptive scenario simulations of thermal performance in semi-open outdoor spaces. Ultimately, these future methodological enhancements will facilitate more intelligent, efficient, and scalable strategies for climate-responsive architectural design.

Author Contributions: Conceptualization, X.M.; Validation, F.Y. and Y.L. (Yuyang Lu); Investigation, H.D.; Resources, Y.Y. and M.Z.; Writing—original draft, Q.L., Y.L. (Yibo Lei), H.C. and H.F.; Funding acquisition, J.Z. All authors have read and agreed to the published version of the manuscript.

Funding: This study was supported by the Xi'an Association for Science and Technology Youth Talent Promotion Project (959202313093), National Natural Science Foundation of China (52278087), Beilin District Science and Technology Project (GX2454) Natural Science Foundation Youth Project of Hubei (Grant No. 2023AFB510), and Weiyang District Science and Technology Plan(202430), Central Universities Basic Scientific Research Business Expenses Special High-tech Research and Cultivation Project (300102415201), Shaanxi Provincial Science and Technology Plan Project (2025JC-YBMS-372) and Natural Science Foundation Youth Project of Hubei (Grant No. 2023AFB510).

Data Availability Statement: The original contributions presented in the study are included in the article, further inquiries can be directed to the corresponding author.

Conflicts of Interest: The authors declare no conflict of interest.

References

1. Lyu, J.; Yang, Y.; Lai, D.; Lai, D.; Lan, L.; Zhao, Z.; Du, H.; Lian, Z. Analysis of occupant thermal comfort and energy-saving potential based on cooling behaviors in residential buildings: A case study of Shanghai. *Build. Environ.* **2025**, *275*, 112792. [CrossRef]
2. Ma, X.; Fukuda, H.; Zhou, D.; Gao, W.; Wang, M. The study on outdoor pedestrian thermal comfort in blocks: A case study of the Dao He Old Block in hot-summer and cold-winter area of southern China. *Sol. Energy* **2019**, *179*, 210–225. [CrossRef]

3. Xu, D.; Zhou, D.; Wang, Y.; Xu, W.; Yang, Y. Field measurement study on the impacts of urban spatial indicators on urban climate in a Chinese basin and static-wind city. *Build. Environ.* **2019**, *147*, 482–494. [CrossRef]
4. Liu, L.; Zhang, Z.; Lan, S.; Tian, X.; Liu, J.; Liao, W.; Wang, D. Developing a spatial optimization design approach towards energy-saving and outdoor thermal comfortable densely-built residential blocks using a dynamic local energy balance model. *Energy Build.* **2025**, *328*, 115194. [CrossRef]
5. Tartarini, F.; Schiavon, S. Comparative analysis of PMV Models accuracy implemented in the ISO 7730: 2005 and ASHRAE 55: 2023. *Build. Environ.* **2025**, *275*, 112766. [CrossRef]
6. Cheung, P.K.; Jim, C. Determination and application of outdoor thermal benchmarks. *Build. Environ.* **2017**, *123*, 333–350. [CrossRef]
7. Potchter, O.; Cohen, P.; Lin, T.P.; Matzarakis, A. Outdoor human thermal perception in various climates: A comprehensive review of approaches, methods and quantification. *Sci. Total Environ.* **2018**, *631–632*, 390–406. [CrossRef] [PubMed]
8. Gagge, A.P.; Fobelets, A.P.; Berglund, L. A standard predictive index of human response to the thermal environment. *Ashrae Trans.* **1986**, *92*, 709–731.
9. Fanger, P.O. *Thermal Comfort*; McGraw Hill: New York, NY, USA, 1970.
10. Jendritzky, G.; de Dear, R.; Havenith, G. UTCI–Why another thermal index? *Int. J. Biometeorol.* **2012**, *56*, 421–428. [CrossRef]
11. Höppe, P. The physiological equivalent temperature—A universal index for the biometeorological assessment of the thermal environment. *Int. J. Biometeorol.* **1999**, *43*, 71–75. [CrossRef]
12. Höppe, P.R. Heat balance modelling. *Experientia* **1993**, *49*, 741–746. [CrossRef]
13. Jamei, E.; Rajagopalan, P. Urban development and pedestrian thermal comfort in Melbourne. *Sol. Energy* **2017**, *144*, 681–698. [CrossRef]
14. Huo, K.; Qin, R.; Zhao, J.; Ma, X. Long-term tracking of urban structure and analysis of its impact on urban heat stress: A case study of Xi'an, China. *Ecol. Indic.* **2025**, *174*, 113418. [CrossRef]
15. Sánchez-Montes, J.; Flores-Prieto, J.; López-Pérez, L.; Ríos-Rojas, C. Adaptive thermal comfort models comparison in dry and rainy seasons: A tropical climate case. *Energy Build.* **2025**, *331*, 115382. [CrossRef]
16. Ma, X.; Fukuda, H.; Zhou, D.; Wang, M. The evaluation of outdoor thermal sensation and outdoor energy efficiency of a commercial pedestrianized zone. *Energies* **2019**, *12*, 1324. [CrossRef]
17. Shashua-Bar, L.; Hoffman, M.E. Vegetation as a climatic component in the design of an urban street: An empirical model for predicting the cooling effect of urban green areas with trees. *Energy Build.* **2000**, *31*, 221–235. [CrossRef]
18. Yang, W.; Wong, N.H.; Li, C.-Q. Effect of street design on outdoor thermal comfort in an urban street in Singapore. *J. Urban Plann. Dev.* **2016**, *142*, 05015003. [CrossRef]
19. Ma, X.; Zhang, L.; Guo, M.; Zhao, J. The effect of various urban design parameter in alleviating urban heat island and improving thermal health—A case study in a built pedestrianized block of China. *Environ. Sci. Pollut. Res.* **2021**, *28*, 38406–38425. [CrossRef]
20. Rosso, F.; Golasi, I.; Castaldo, V.L.; Piselli, C.; Pisello, A.L.; Salata, F.; Ferrero, M.; Cotana, F.; de Lieto Vollaro, A. On the impact of innovative materials on outdoor thermal comfort of pedestrians in historical urban canyons. *Renew. Energy* **2018**, *118*, 825–839. [CrossRef]
21. Ma, X.; Wang, M.; Zhao, J.; Zhang, L.; Liu, W. Performance of Different Urban Design Parameters in Improving Outdoor Thermal Comfort and Health in a Pedestrianized Zone. *Int. J. Environ. Res. Public Health* **2020**, *17*, 2258. [CrossRef]
22. Zhang, L.; Ma, X.; Zhao, J.; Wang, M. Tourists' Thermal Experience and Health in a Commercial Pedestrianized Block: A Case Study in a Hot and Humid Region of Southern China. *Int. J. Environ. Res. Public Health* **2019**, *16*, 5072. [CrossRef] [PubMed]
23. Rosso, F.; Pisello, A.L.; Cotana, F.; Ferrero, M. On the thermal and visual pedestrians' perception about cool natural stones for urban paving: A field survey in summer conditions. *Build. Environ.* **2016**, *107*, 198–214. [CrossRef]
24. Thorsson, S.; Lindqvist, M.; Lindqvist, S. Thermal bioclimatic conditions and patterns of behaviour in an urban park in Goteborg, Sweden. *Int. J. Biometeorol.* **2004**, *48*, 149–156. [CrossRef]
25. Wang, Y.; de Groot, R.; Bakker, F.; Wörtche, H.; Leemans, R. Thermal comfort in urban green spaces: A survey on a Dutch university campus. *Int. J. Biometeorol.* **2017**, *61*, 87–101. [CrossRef] [PubMed]
26. Yang, B.; Olofsson, T.; Nair, G.; Kabanshi, K. Outdoor thermal comfort under subarctic climate of north Sweden—A pilot study in Umeå. *Sustain. Cities Soc.* **2017**, *28*, 387–397. [CrossRef]
27. He, Y.; Zhong, H.; Fan, Y.; Lau, S.S.Y.; Cui, G. A calibration framework for outdoor thermal comfort index and neutral range: Application in subtropical hot-humid campus. *Indoor Built Environ.* **2025**, *34*, 554–568. [CrossRef]
28. Chen, L.; Wen, Y.Y.; Zhang, L.; Xiang, W.N. Studies of thermal comfort and space use in an urban park square in cool and cold seasons in Shanghai. *Build. Environ.* **2015**, *94*, 644–653. [CrossRef]
29. Liu, W.; Zhang, Y.; Deng, Q. The effects of urban microclimate on outdoor thermal sensation and neutral temperature in hot-summer and cold-winter climate. *Energy Build.* **2016**, *128*, 190–197. [CrossRef]
30. Zhang, L.; Zhan, Q.; Lan, Y. Effects of the tree distribution and species on outdoor environment conditions in a hot summer and cold winter zone: A case study in Wuhan residential quarters. *Build. Environ.* **2018**, *130*, 27–39. [CrossRef]

31. Li, K.; Zhang, Y.; Zhao, L. Outdoor thermal comfort and activities in the urban residential community in a humid subtropical area of China. *Energy Build.* **2016**, *133*, 498–511. [CrossRef]
32. Irakoze, A.; Kim, J.; Kim, K.H. Development of a real time setpoint temperature control method based on artificial neural network (ANN) and adaptive thermal comfort. *Energy Build.* **2025**, *331*, 115390. [CrossRef]
33. Fang, Z.; Feng, X.; Liu, J.; Lin, Z.; Mak, C.M.; Niu, J.; Tse, K.T.; Xu, X. Investigation into the differences among several outdoor thermal comfort indices against field survey in subtropics. *Sustain. Cities Soc.* **2019**, *44*, 676–690. [CrossRef]
34. Huang, J.; Zhou, C.; Zhuo, Y.; Xu, L.; Jiang, Y. Outdoor thermal environments and activities in open space: An experiment study in humid subtropical climates. *Build. Environ.* **2016**, *103*, 238–249. [CrossRef]
35. Li, J.; Niu, J.; Mak, C.M.; Huang, T.; Xie, Y. Assessment of outdoor thermal comfort in Hong Kong based on the individual desirability and acceptability of sun and wind conditions. *Build. Environ.* **2018**, *145*, 50–61. [CrossRef]
36. Chen, X.; Xue, P.; Liu, L.; Gao, L.; Liu, J. Outdoor thermal comfort and adaptation in severe cold area: A longitudinal survey in Harbin, China. *Build. Environ.* **2018**, *143*, 548–560. [CrossRef]
37. Shui, T.; Liu, J.; Yuan, Q.; Qu, Y.; Jin, H.; Cao, J.; Liu, L.; Chen, X. Assessment of pedestrian–level wind conditions in severe cold regions of China. *Build. Environ.* **2018**, *135*, 53–67. [CrossRef]
38. Chen, X.; Xue, P.; Gao, L.; Du, J.; Liu, J. Physiological and thermal response to real—Life transient conditions during winter in severe cold area. *Build. Environ.* **2019**, *157*, 284–296. [CrossRef]
39. *GB 50176–2016*; Thermal Design Code for Civil Buildings, TDCFCB. China Architecture & Building Press: Beijing, China, 2016.
40. Lai, D.; Guo, D.; Hou, Y.; Lin, C.; Chen, Q. Studies of outdoor thermal comfort in northern China. *Build. Environ.* **2014**, *77*, 110–118. [CrossRef]
41. Yan, H.; Wu, F.; Dong, L. Influence of a large urban park on the local urban thermal environment. *Sci. Total Environ.* **2018**, *622–623*, 882–891. [CrossRef]
42. Xu, M.; Hong, B.; Mi, J.; Yan, S. Outdoor thermal comfort in an urban park during winter in cold regions of China. *Sustain. Cities Soc.* **2018**, *43*, 208–220. [CrossRef]
43. Lin, T.P. Thermal perception, adaptation and attendance in a public square in hot and humid regions. *Build. Environ.* **2009**, *44*, 2017–2026. [CrossRef]
44. Wang, Y.; Bakker, F.; de Groot, R.; Wörtche, H.; Leemans, R. Effects of urban green infrastructure (UGI) on local outdoor microclimate during the growing season. *Monit. Assess.* **2015**, *187*, 732. [CrossRef] [PubMed]
45. Yin, J.; Zheng, Y.; Tan, J.; Ye, D.; Wang, W. An analysis of influential factors on outdoor thermal comfort in summer. *Int. J. Biometeorol.* **2012**, *56*, 941–948. [CrossRef]
46. Liu, J.; Niu, J.; Xia, Q. Combining measured thermal parameters and simulated wind velocity to predict outdoor thermal comfort. *Build. Environ.* **2016**, *105*, 185–197. [CrossRef]
47. Niu, J.; Liu, J.; Lee, T.-C.; Lin, Z.; Mak, C.; Tse, K.-T.; Tang, B.-S.; Kwok, K.C. A new method to assess spatial variations of outdoor thermal comfort: Onsite monitoring results and implications for precinct planning. *Build. Environ.* **2015**, *91*, 263–270. [CrossRef]
48. Liu, J.; Niu, J. CFD simulation of the wind environment around an isolated high rise building: An evaluation of SRANS, LES and DES models. *Build. Environ.* **2016**, *96*, 91–106. [CrossRef]
49. Xia, Q.; Liu, X.; Niu, J.; Kwok, K.C.S. Effects of building lift-up design on the wind environment for pedestrians. *Indoor Built Environ.* **2015**, *26*, 1214–1231. [CrossRef]
50. Wang, D.; Xu, H.; Yang, J.; Luo, Y.; Yang, L. Energy consumption analysis and electricity pricing strategies of university dormitories in hot summer and warm winter region: A case from Guangzhou University, South China. *Energy Build.* **2025**, *335*, 115525. [CrossRef]
51. Peel, M.C.; Finlayson, B.L.; Mcmahon, T.A. Updated world map of the Köppen—Geiger climate classification. *Hydrol. Earth Syst. Sci.* **2007**, *11*, 259–263. [CrossRef]
52. Meteorological Organization Country. Available online: https://www.cma.gov.cn/ (accessed on 31 July 2025).
53. Meteorological Organization of Xi'an City. Available online: http://sn.cma.gov.cn/dsqx/xaqx/ (accessed on 31 July 2025).
54. Alfano, F.R.D.; Dell'iSola, M.; Ficco, G.; Palella, B.I.; Riccio, G. Small globes and pocket heat stress meters for WBGT and PHS evaluations. A critical analysis under controlled conditions. *Build. Environ.* **2022**, *226*, 109781. [CrossRef]
55. Bruse, M. ENVI-Met 4.4: Updated Model Overview, University of Bochum. 2021. Available online: http://www.envi-met.com (accessed on 31 July 2025).
56. Jiang, Y.; Han, X.; Shi, T.; Song, D. Adaptive analysis of green space network planning for the cooling effect of residential blocks in summer: A case study in shanghai. *Sustainability* **2018**, *10*, 3189. [CrossRef]
57. Zheng, B.; Bedra, K.B.; Zheng, J.; Wang, G. Combination of tree configuration with street configuration for thermal comfort optimization under extreme summer conditions in the urban center of Shantou City. *Sustainability* **2018**, *10*, 4192. [CrossRef]
58. Bochenek, A.D.; Klemm, K. The impact of passive green technologies on the microclimate of historic urban structures: The case study of lodz. *Atmosphere* **2020**, *11*, 974. [CrossRef]
59. Willmott, C.J. Some Comments on the Evaluation of Model Performance. *Bull. Am. Meteorol. Soc.* **1982**, *63*, 1309–1313. [CrossRef]

60. *ISO7726*; Ergonomics of the Thermal Environment-Instruments for Measuring Physical Quantities. International Standards Organization: Geneva, Switzerland, 2003.
61. Mi, J.; Hong, B.; Zhang, T.; Huang, B.; Niu, J. Outdoor thermal benchmarks and their application to climateresponsive designs of residential open spaces in a cold region of China. *Build. Environ.* **2020**, *169*, 106592. [CrossRef]
62. *GB 50352-2019*; Uniform Standard for Design of Civil Buildings. National Standard of the People's Republic of China: Beijing, China, 2019. Available online: https://baijiahao.baidu.com/s?id=1668633499138401533&wfr=spider&for=pc (accessed on 4 August 2023).
63. Liu, L.; Xia, B.; Zhang, L. Outdoor thermal comfort of urban river landscape belt in China's cold region: A case study of Xi'an. *Urban Clim.* **2023**, *48*, 101406. [CrossRef]
64. Yin, P.; Ji, Y.; Xie, J.; Liu, J.; Hou, Q.; Zhao, S.; Jing, P. Residential wintry thermal comfort and adaptive behaviors in a cold climate in Beijing, China. *Energy Build.* **2022**, *265*, 111942. [CrossRef]

buildings

MDPI

Article

Optimal Spatial Configuration for Energy and Solar Use in Alpine-Frigid Resettlement Communities

Bo Liu [1,†], Wei Song [2,†], Yu Liu [1,3,4,*], Chuanming Wang [2] and Jie Song [4,5]

1 School of Architecture, Xi'an University of Architecture and Technology, Xi'an 710055, China; liubo@xauat.edu.cn
2 Power China Northwest Engineering Corporation Limited, Xi'an 710065, China
3 School of Mechanics and Transportation Engineering, Northwestern Polytechnical University, Xi'an 710129, China
4 Sustainable Building and Environmental Research Institute, Northwestern Polytechnical University, Xi'an 710129, China
5 School of Mechanical Engineering, Northwestern Polytechnical University, Xi'an 710129, China
* Correspondence: liuyu@nwpu.edu.cn; Tel.:+86-139-9116-4757
† These authors contributed equally to this work.

Abstract: Resettlement communities in Qinghai are located in cold, high-altitude regions with dry climates and strong solar radiation. Although not extremely cold, the moderate heating demand aligns well with high solar availability, making passive design highly effective for reducing energy use. This study investigates solar-optimized spatial configurations that enhance passive energy performance while addressing functional settlement needs. Through parametric modeling and climate-responsive simulations, four key spatial parameters are examined: building spacing, courtyard depth, density, and volumetric ratio. The findings highlight the dominant role of front–rear spacing in solar access, with optimal values at 3–4 m for single-story and 5–10 m for two-story buildings, balancing radiation gain and land use efficiency. Courtyard depths under 2.7 m significantly limit south façade exposure due to shading from the opposite courtyard wall under low-angle winter sun. This reduction results in the south façade attaining only 55.7–79.6% of the solar radiation acquisition by an unobstructed south façade (the baseline). Meanwhile, clustered orientations reduce inter-building shading losses by 38–42% compared to dispersed layouts. A three-tiered design framework is proposed: (1) macro-scale solar orientation zoning, (2) meso-scale spacing tailored to building height, and (3) micro-scale courtyard modulation for low-angle winter radiation. Together, these strategies provide practical, scalable guidelines for energy-efficient, climate-responsive settlement design in the alpine regions of Qinghai.

Keywords: alpine resettlement communities; building layout optimization; passive energy efficiency; solar energy utilization; spatial configuration

1. Introduction

1.1. Background

As pivotal hubs for hydropower development in China, Qinghai hosts a significant number of resettlement communities for hydropower project migrants. These projects must pursue sustainable pathways within fragile, high-altitude arid ecosystems [1]. The region exhibits a distinctive "high solar radiation–moderate-to-low heating energy consumption (HEC)" profile: The combination of intense solar exposure and drastic diurnal/seasonal temperature fluctuations creates exceptional potential for passive solar design—abundant solar resources align well with relatively modest heating demands in timing and intensity. Nevertheless, winters impose substantial heating loads, posing critical challenges for settlement planning.

While significant progress has been made in building energy efficiency research for cold high-altitude regions, a scale-related limitation persists, as existing studies predominantly concentrate on optimizing materials, building envelopes, and active technologies at the individual building level. Research on how community-scale spatial configurations systematically influence solar radiation acquisition and HEC remains underexplored. This gap similarly exists in the standardized planning protocols for resettlement communities, which often fail to adequately incorporate site-specific solar accessibility analysis and its spatial variation across building clusters, leading to suboptimal daylight utilization.

Spatial morphological parameters constitute fundamental variables governing solar radiation acquisition. They directly determine effective solar irradiance on building surfaces, shadowing effects, passive heat gain potential, and heat loss rates, thereby critically impacting HEC and overall energy performance. Despite recognition of their importance, the specific quantitative mechanisms through which village-scale spatial morphology regulates the complete "solar gain–heat loss" chain and ultimately governs community-level HEC require systematic investigation and quantification.

Consequently, this study focuses on hydropower resettlement communities and similar alpine villages in Qinghai. It aims to elucidate the integrated regulatory mechanisms by which village spatial form parameters affect solar radiation acquisition efficiency and heat loss. By exploring optimized spatial arrangements, the research pursues the dual goals of enhancing solar gain and reducing heat loss. The findings provide evidence-based design guidelines and technical support for developing energy-efficient settlements in high-altitude cold regions.

1.2. Literature Review

The interplay between urban morphology and solar radiation accessibility has garnered significant scholarly attention in recent years. Liu et al. [2] systematically reviewed 258 studies (2011–2022) to evaluate 111 urban form indicators, revealing that building density (BD) (total building footprint area/site area) and floor area ratio (FAR) (total floor area/site area) exhibit negative correlations with solar radiation acquisition, whereas building height and window-to-wall ratio demonstrate positive associations. Their analysis emphasized boundary condition dependencies and identified substitutive relationships among specific morphological metrics, providing a theoretical foundation for future urban energy modeling. Addressing latitude-specific challenges, Xiao et al. [3] developed a predictive model integrating building morphology parameters with latitude cosine values, achieving high accuracy in estimating residential block solar radiation (adjusted R^2: 0.931 for roofs, 0.962 for façades), thereby offering a versatile tool for multi-climate applications.

Manni et al. [4] developed a solar potential mapping framework tailored for high-latitude winter conditions with low irradiance in Trondheim, Norway, by integrating LiDAR-derived 3D models with radiation simulations. Freitas et al. [5] systematically analyzed the impact mechanisms of vertical expansion and dynamic shading effects on solar potential assessment, revealing methodological trade-offs inherent in relevant digital tools. Gan et al. [6] employed a simulation-based genetic algorithm to optimize Hong Kong high-rise residential layouts, achieving 30–40% reductions in air conditioning and lighting energy by maximizing natural ventilation and daylight.

He et al. [7] simulated 384 Hong Kong cases, showing that building height heterogeneity reduces net carbon emissions by minimizing shading, while shape heterogeneity (footprint aspect ratio/surface-to-volume ratio) increases energy penalties despite enhanced facade solar gains. Dino et al. [8] developed the multi-objective tool MADE. It uses a two-step genetic algorithm—first resolving spatial layout constraints, then co-optimizing energy and daylighting—to generate Pareto-optimal building designs, validated via a library case study. Hachem-Vermette et al. [9] found that in their Calgary prototypes, optimized orientations resulted in less than 3% solar access variation across layouts. However, hexagonal networks outperformed rectangular configurations on resilience metrics. Asfour et al. [10] confirmed that horizontally compact layouts in tropical regions achieve 28% greater energy savings compared to vertical forms. Kosir et al. [11] analyzed seven typical Slovenian urban layouts, revealing that solar potential constraints are less severe than expected, with building form/orientation posing the primary challenge.

Deng et al. [12] verified in Harbin that courtyard layouts optimized for spacing and orientation can yield 4~5% energy savings. Xu et al. [13] proposed an evaluation workflow for Wuhan office blocks. It revealed that combined morphological parameters significantly impact energy use intensity (cooling: 28.83%, heating: 28.56%, lighting: 23.23%), identifying building shape factor and FAR as dominant factors. Ahmed et al. [14] utilized building performance simulation to drive a 51.3% energy reduction target for Saudi residential buildings. Yan et al. [15] proposed using solar radiation intensity as a quantitative shading indicator in hot summer/cold winter zones. Optimizing residential cluster layouts achieved 4.39% cooling and 4.03% HEC reductions at the cluster scale. Narimani Abar et al. [16] identified 10 urban morphological variables influencing residential energy consumption. Their predictive model (~80% accuracy) resolved controversies regarding density's impact: density correlates negatively with energy use in cold climates but potentially positively in temperate zones, contingent upon climate, energy type, and research methodology. A GIS-based urban-scale building energy modeling tool that automates building stock performance assessment, defines energy efficiency measures, introduces an urban building energy rating, and was validated on 769 Italian residential buildings was presented by de Rubeis et al. [17]. Du et al. [18] developed a computational method for optimizing office building space layouts to enhance energy efficiency, identifying the facade-to-FAR as the most influential design indicator; their layout optimization demonstrates potential reductions of up to 54% in lighting, heating, and cooling demands. Zhu et al. [19] investigated the dual impact of urban morphology on building energy consumption and outdoor thermal comfort (UTCI) in hot–humid climates. Simulations of six block layouts reveal that parameters like FAR and plan dimensions universally influence performance, while surface-to-volume ratio (S/V) and sky view factor (SVF) exhibit sensitivities dependent on specific layout configurations.

Hadavi et al. [20] demonstrated, in their Tehran-based study, that reduced urban compactness lowers urban heat island (UHI) intensity and saves 16.4% energy through thermal wake evacuation, while rooftop units experience overheating due to 50K surface-window temperature gaps. Wu et al. [21] quantified the dual impact of urban morphology on building energy consumption and renewable energy potential at the neighborhood scale in Beijing. They revealed the interplay between the UHI effect (reducing heating energy by 15.8% but increasing cooling energy by 30%) and building shading effects (increasing heating energy by 11.88% but decreasing cooling energy by 5.87%), providing morphological parameters for precise renewable energy allocation.

Hong et al. [22], through numerical simulations, show that building layouts with long facades parallel to prevailing winds accelerate horizontal vortices, improving pedestrian-level thermal comfort. Furthermore, configurations featuring central squares oriented towards the wind direction, combined with strategic tree placement, optimize the microclimate by reducing standard effective temperature (SET) values. Yan et al. [23] addressed the severe 6.5% annual increase in building energy consumption in hot summer/cold winter zones. By quantifying the impact of residential cluster layouts on energy and carbon emissions using an effective shading coefficient, they revealed optimization paths; staggered layouts reduce cooling energy by 7.17%, and enclosed layouts reduce heating energy by 4.06%, offering key emission reduction strategies for high-growth regions.

Lu et al. [24] identified three high-rise residential layout patterns in Northeast China via SPSS (SPSS version 29) multivariate response analysis. Combined with Ecotect radiation simulations, they revealed that three-sided enclosed layouts offer optimal solar potential, with specific nonlinear relationships between FAR, BD, height, and radiation, providing quantitative guidelines for high-density solar optimization.

Singh et al. [25] proposed a three-step workflow for community-scale net-zero energy retrofits in Canada; envelope upgrades can reduce energy use by 60%. The optimal combination of solar strategies is highly dependent on community type and street layout, providing a systematic decision-making tool for climate-adaptive community planning. Sosa et al. [26] addressed social housing communities in arid regions. Microclimate simulations revealed that optimizing street layout/orientation combined with street trees + high-albedo materials can reduce summer cooling energy consumption by over 21%, providing a thermal environment regulation paradigm for high-temperature urban communities. Huo et al. [27] utilized remote sensing and mesoscale modeling to track Xi'an's urban structure, finding that its evolution towards taller, less dense buildings correlated with reduced heat stress.

1.3. Research Gap and Contribution

While robust linkages between urban morphology, energy efficiency, and solar potential are established across diverse climates, high-altitude resettlement communities—characterized by prevailing single-/double-story dwellings—remain under-investigated. Critical unmet needs include: (1) layout optimization for dominant low-rise typologies, (2) collective shading dynamics in clustered developments, (3) cold-climate solar strategies for courtyards and walls, and (4) integrated frameworks balancing solar gain with land-use efficiency.

Theoretically, this study advances bioclimatic design theory through quantification of cluster morphology impacts on solar radiation acquisition and HEC in alpine settlements. Mechanistic insights into building energy flow modulation by collective shading are revealed via integration of radiative heat transfer dynamics with spatial parameter analysis, extending climate-responsive principles to address spatial energy interdependencies in vulnerable resettlement contexts.

Practically, transferable protocols are developed, including cluster-scale spatial standards enabling low-carbon settlement design and passive heating strategies tailored to cold regions. Evidence-based tools are thereby yielded that harmonize energy efficiency with ecological preservation in high-altitude development.

1.4. Research Purpose

The dual dimensions of building cluster spatial morphology—exterior surface solar radiation acquisition and building HEC—are investigated in hydropower resettlement communities across cold high-altitude regions. Critical morphological parameters (orientation, density, courtyard design) are quantitatively analyzed to develop climate-responsive strategies that concurrently maximize passive solar radiation acquisition and minimize HEC. Significant reductions in HEC are targeted through this integrated framework, while thermal comfort improvements are enhanced. Actionable guidelines for sustainable settlement planning in high-altitude and cold environments are thereby formulated.

2. Methodology

2.1. Research Path and Method

A combined methodology of parametric simulation and quantitative trend analysis is employed to systematically examine the coupling mechanisms between spatial morphology, exterior surface solar radiation acquisition, and building HEC in resettlement communities across Qinghai. Georeferenced 3D parametric models are developed in Rhino 8 using field-surveyed data, with core morphological variables—building spacing, orientation, courtyard configuration, and density—explicitly defined. Plateau-specific meteorological data are integrated to drive simulations.

Climate-responsive workflows are implemented through Ladybug Tools 1.8.0 coupled with EnergyPlus 25.1.0. Exterior surface solar radiation acquisition and HEC are quantified across morphological scenarios. The controlled variable method is applied to analyze impact trends of individual parameters. Characteristic features of parameter-performance curves (e.g., growth inflection points, fluctuation thresholds) and performance variation magnitudes are observed to determine directional influences and effect intensities of morphological elements. Spatial design guidelines prioritizing key interventions are subsequently distilled for enhanced solar radiation acquisition.

This integrated modeling-simulation-trend analysis framework establishes explicit correlations between cold-region settlement morphology and energy performance, providing theoretical foundations for high-altitude habitat planning.

2.2. Variable Identification

The methodology systematically defines morphological drivers and energy response indicators to precisely examine spatial energy interactions. Four core layout parameters are established as independent variables: building spacing (front–rear and side distances, simulated range: 0 m–20 m), courtyard depth (radiative accessibility, simulated range: 0.9 m–6 m), BD (footprint ratio, controlled by building spacing), and FAR (spatial compactness, controlled by building spacing and number of stories). Dependent variables quantify radiation performance (annual cumulative surface irradiation) and thermal demand (heating load intensity). Building height, orientation, and simplified rectangular morphology are standardized as fixed parameters, while meteorological inputs are anchored to the calibrated EnergyPlus Weather Data for Xining, Qinghai (https://energyplus.net/weather, (accessed on 5 May 2025)). This controlled parameter system enables precise analysis of morphology–energy pathways in alpine environments.

2.3. Sample and Data Sources

An idealized prototype model is developed based on typical resettlement patterns, with real-world complexity abstracted into parametric spatial grids. Critically, this methodological approach aligns with the standard [28], wherein prescribed regional heating energy limits are derived from identical idealized simulations, confirming the validity of our abstraction framework. Orthogonal building volumes with uniform height and north–south orientation eliminate geometric interference, while thermal parameters are rigorously set according to standard [28,29]. The computational workflow is fully automated within the Rhinoceros-Grasshopper environment through embedded physics engines (Ladybug/Honeybee), eliminating manual formula implementation, as follows:

1. Solar flux mapping: Hourly irradiation simulations generate surface radiation heat flux profiles.
2. Thermal load calculation: Steady-state heat transfer models quantify envelope heat loss.

3. Performance benchmarking: Energy efficiency differentials are assessed against baseline scenarios.

This abstraction strategy focuses on spatial parameter interactions to enhance analytical resolution, generating transferable planning principles for cold arid settlements. Modular technical architecture maintains computational feasibility while enabling seamless scalability to practical resettlement projects.

3. Results and Discussion

3.1. Climatic Context

Qinghai demonstrates a stark bioclimatic duality defined by the coexistence of intense solar radiation and prolonged high-altitude cold arid conditions. Marked seasonal temperature extremes prevail, featuring extended sub-zero winters contrasting with dramatic daily summer temperature fluctuations in southern regions. Chronically low humidity significantly intensifies radiative cooling effects throughout the province.

This paradox necessitates dual energy strategies: severe heating demands are counterbalanced by exceptional solar abundance, positioning passive solar optimization as critical for sustainable settlements. Altitude-modulated climates impose non-negotiable design imperatives—maximization of radiative heat gain during extended winters must be systematically reconciled with mitigation of seasonal overheating risks. Such contradictory requirements fundamentally redefine spatial planning protocols for resettlement communities across energy-critical high-altitude zones.

3.2. Validation Protocol

Validation benchmarks were established using a simplified rectangular prototype (Figure 1) calibrated to Qinghai's construction standards. The 2:1 plan ratio reflects prevalent geometries in surveyed settlements, and the absence of north fenestration is inherent to the single-room abstraction where south-oriented windows suffice for daylighting. Window thermal parameters were optimized through U-value/SHGC adjustments for alpine radiative balance. Simulated heating loads aligned with standards [28,29] within 2% error, confirming model fidelity in altitude-specific heat transfer analysis.

Figure 1. Verification model.

3.3. Impacts of Building Spacing on Solar Radiation Acquisition and Heating Energy Consumption

This study evaluates shading effects by comparing radiation values of obstructed configurations against an unobstructed baseline (Figures 2 and 3). Simulations reveal a nonlinear inverse correlation between building spacing and radiation loss, where shading impact becomes negligible once spacing exceeds a critical threshold. North–south building rows require larger spacing than east–west orientations to mitigate shading.

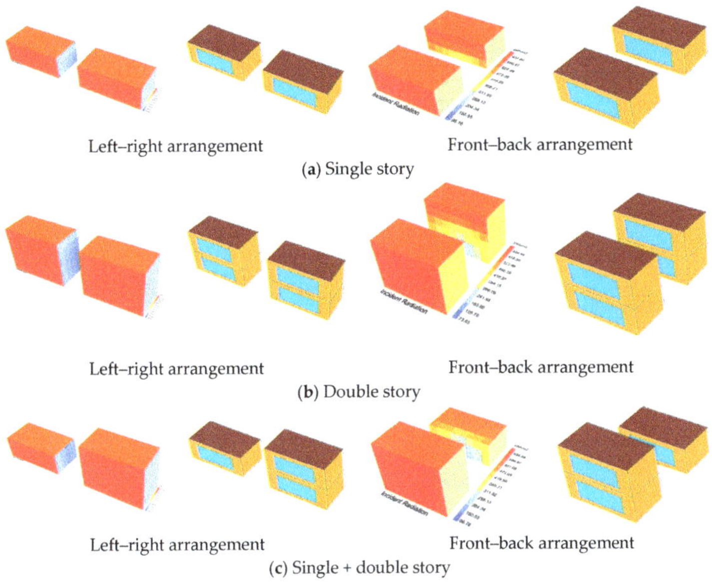

Figure 2. Solar radiation and HEC model of building arrangement.

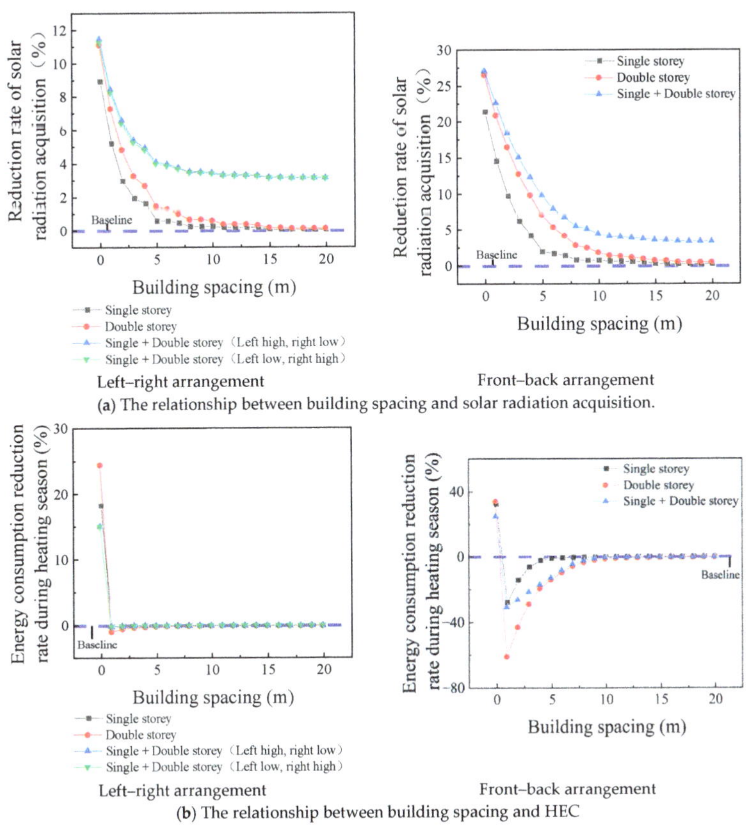

Figure 3. The influence of building spacing on solar radiation acquisition and building HEC.

Thermal analysis demonstrates that terraced layouts reduce heating demand by 18% (single-layer), 24% (double-layer), and 15% (mixed-layer) compared to detached layouts through minimized external surface exposure (Figure 3), confirming their efficacy in balancing energy efficiency and solar utilization. Front–back spacing optimization shows diminishing thermal returns beyond 5 m (uniform height units) and 10 m (hybrid configurations)—thresholds consistent with observed radiation equilibria (Table 1).

Table 1. Critical building spacing at curve leveling-off point of solar radiation reduction rate under different boundary conditions.

Condition	Critical Building Pacing at Leveling-Off Point
Single story, left–right arrangement	5 m
Single story, front–back arrangement	5 m
Double story, left–right arrangement	5 m
Double story, front–back arrangement	10 m
Single + double story, left–right (left high)	5 m
Single + double story, left–right (left low)	5 m
Single + double story, front–back (front high, back low)	10 m
Single story, left–right arrangement	5 m

3.4. Impacts of Building Orientation on Solar Radiation Acquisition and Heating Energy Consumption

A significant directional effect was revealed through the establishment of rotational scenarios for both individual buildings and building clusters (Figure 4). At the individual building scale, maximum solar radiation acquisition was achieved at a 10° west-of-south orientation, whereas the minimum HEC was observed at 60° east-of-south. A 5% differential in HEC was recorded between these orientations during the heating season, indicating that angular deviation from the 10° west-of-south orientation should be minimized.

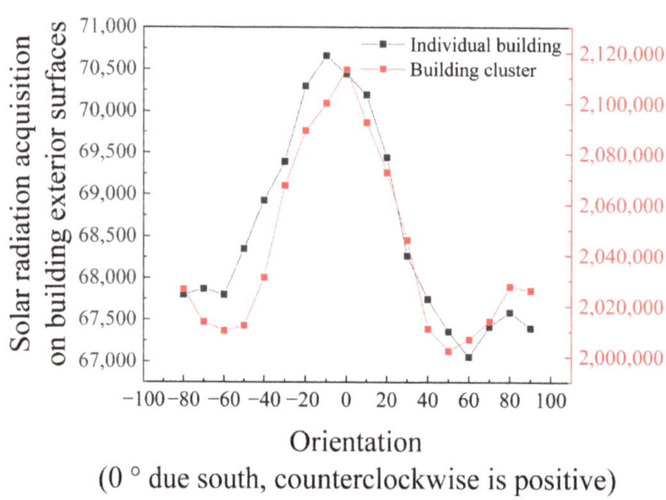

Figure 4. The impact of building orientation changes on the solar radiation acquisition by the building's exterior surfaces.

For building clusters, due south alignment was found to maximize mutual shading reduction, establishing it as the optimal village layout pattern. By contrast, a 50° east-of-south cluster orientation resulted in approximately 5% higher energy loads. This discrepancy necessitates differentiated design strategies: the 10° west-of-south orientation is prioritized for individual structures to enhance solar capture efficiency, while due south alignment is maintained for building groups to optimize collective radiation efficiency.

The thermal performance analysis (Figure 5) reveals the distinct optimization strategies required for building orientation at different spatial scales. For individual buildings, the optimal orientation is 20° east of south, which maximizes morning solar radiation acquisition and reduces HEC. In contrast, building clusters perform best when oriented 40° east of south, a direction that balances solar radiation distribution within the cluster and minimizes mutual shading between buildings to achieve peak efficiency.

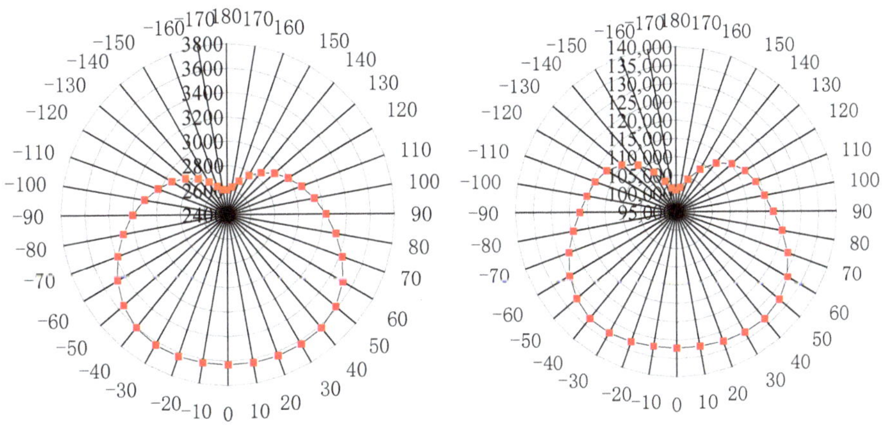

(**a**) Individual building

(**b**) Building cluster

Figure 5. Impact of different building orientations on HEC.

Accordingly, the design guidelines recommend orientation ranges that consider both solar optimization and layout flexibility: from 50° east of south to 40° west of south for individual buildings, and from 60° east of south to 60° west of south for building clusters. These threshold ranges not only help accommodate future climate variability but also effectively mitigate the negative impacts of shading within clusters.

The findings establish a scale-sensitive design paradigm; while individual buildings require precise orientation tuning to optimize solar gain, building clusters must seek radiation synergy within a broader angular tolerance. This dual-strategy framework significantly enhances the environmental adaptability of resettlement housing projects, systematically reduces HEC, and strongly supports regional goals for sustainable development in the building energy sector.

3.5. Impact of Courtyard Depth on Solar Radiation Acquisition and Heating Energy Consumption

This study evaluates the shading effects of courtyards using 2 m high walls with depths ranging from 0.9 m to 6 m (Figure 6). The solar accessibility of the south façade shows a nonlinear negative correlation with courtyard depth, using the unobstructed condition (100% exposure) as a reference. Within the 1–3 m depth range, solar radiation acquisition fluctuates between 55.7% and 79.6%, reflecting significant shading effects. Beyond 3 m, the attenuation rate decreases notably, stabilizing around 86.6% at a depth of 4.8 m, indicating that persistent east–west wall shadows during low solar angles hinder full radiation recovery (Figures 7 and 8).

Figure 6. Radiation model with courtyard walls.

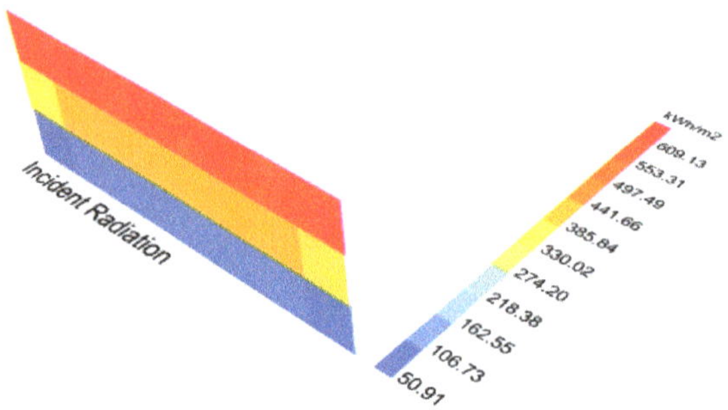

Figure 7. Solar radiation acquisition on the south façade.

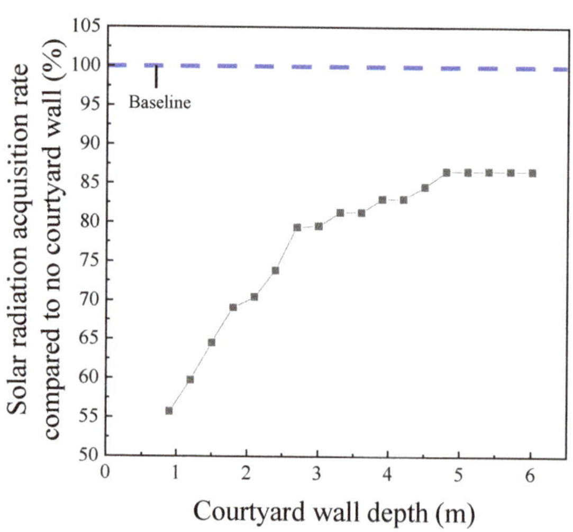

Figure 8. Courtyard wall depth versus south façade solar radiation acquisition rate (100% under unobstructed conditions).

The results identify three operational zones:

1. High-sensitivity zone (0.9–3 m): Approximately 12% reduction in irradiance for every additional meter in depth.
2. Transition zone (3–4.8 m): Marginal losses dominate, with irradiance loss per meter falling below 4%.
3. Stabilization threshold (>4.8 m): A residual 13.4% reduction in irradiance persists due to lateral obstructions.

These nonlinear dynamics necessitate depth-specific design protocols, as follows: for solar-critical applications, courtyard depths should be kept below 3 m; beyond 4.8 m, greater flexibility is allowed due to the diminished impact of shading.

Using the HEC under unobstructed exposure as the baseline (100%), the modulation effect of courtyard configurations (Figure 9) on building thermal load exhibits a clear depth dependency (Figure 10). Within the tested depth range (0.9 m to 6 m), HEC fluctuates between 78.3% and 95.1% of the baseline, revealing a critical nonlinear dynamic.

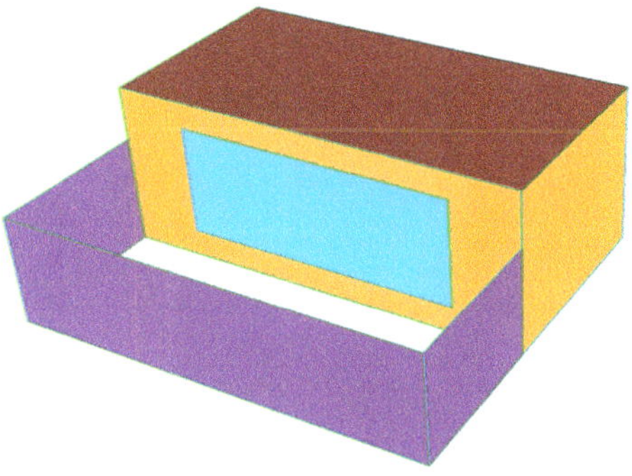

Figure 9. HEC model with courtyard walls.

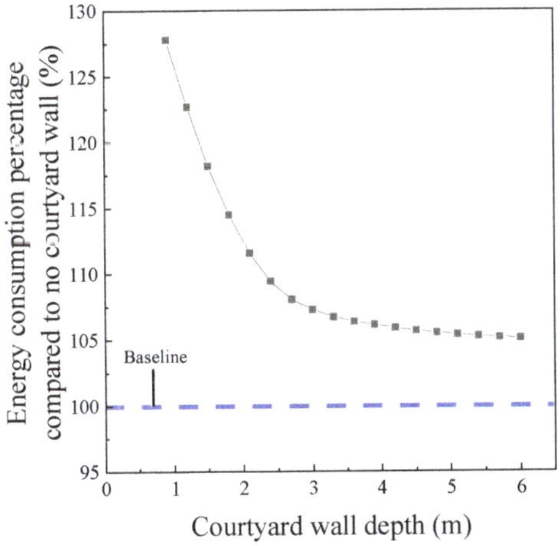

Figure 10. Impact of courtyard wall depth on building HEC percentage (100% under no-wall condition).

When courtyard depth is less than 2.7 m, HEC increases sharply with depth—by as much as 16.8%—due to significant obstruction of solar gain on the south façade, particularly during periods of low winter radiation. However, beyond 2.7 m, the sensitivity of thermal load-to-depth decreases markedly; between 2.7 m and 6 m, HEC shows only a marginal change of 3.2%. This suggests that during low solar altitude periods, the shading effect of east–west courtyard walls reaches a saturation point, beyond which, additional depth does not substantially increase HEC.

Based on this threshold, the study proposes targeted design principles, as follows: courtyard depth should be kept within 2.7 m to minimize winter heating losses; for depths greater than 2.7 m, more spatial flexibility can be allowed without significantly compromising thermal performance.

3.6. Impact of Building Density and Floor Area Ratio on Solar Radiation Acquisition and Heating Energy Consumption

When investigating the impact of BD and FAR on solar radiation acquisition and HEC, this study systematically adjusted building spacing and single- or double-story configurations within fixed site boundary conditions to vary density and FAR parameters (Figure 11). For single-story building units (Figures 12 and 13), the influence of north–south (front-to-back) spacing on solar radiation was more than twice as strong as that of east–west (lateral) spacing. Based on a 2% irradiance variance threshold, the optimal spacing ranges were identified as 1–1.5 m east–west and 3–4 m north–south.

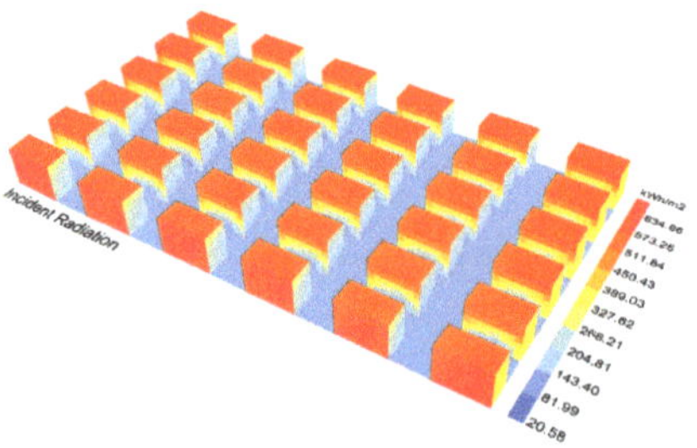

Figure 11. Schematic model of buildings and site boundaries.

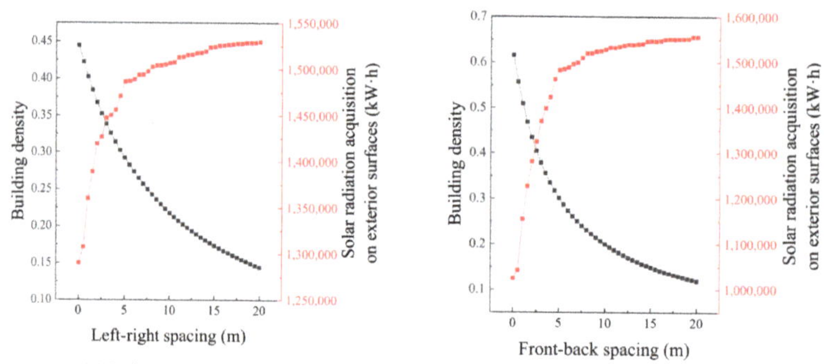

(**a**) Left–right arrangement (**b**) Front–back arrangement

Figure 12. Impact of spacing and BD on solar radiation acquisition for single-story buildings.

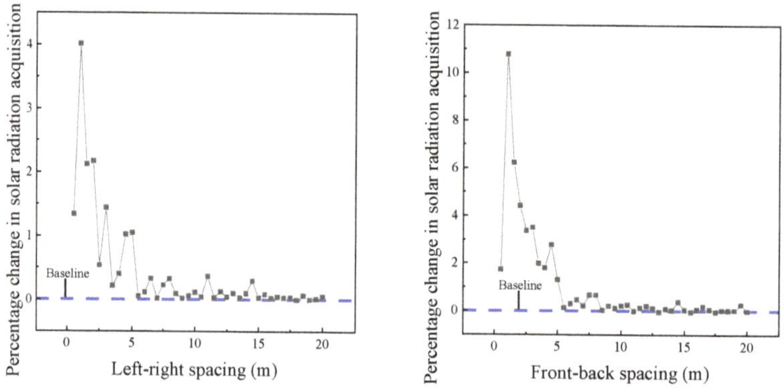

(**a**) Left–right arrangement (**b**) Front–back arrangement

Figure 13. Effect of spacing on percentage change in solar radiation acquisition for single-story buildings.

For double-story building structures (Figures 14 and 15), the increased vertical surface area exposed to sunlight required significantly larger spacing—expanded to 3–5 m east–west and extended to 5–10 m north–south—to ensure adequate solar exposure on higher façades. This hierarchical spacing strategy provides a quantitative design basis to balance development intensity with solar energy utilization efficiency.

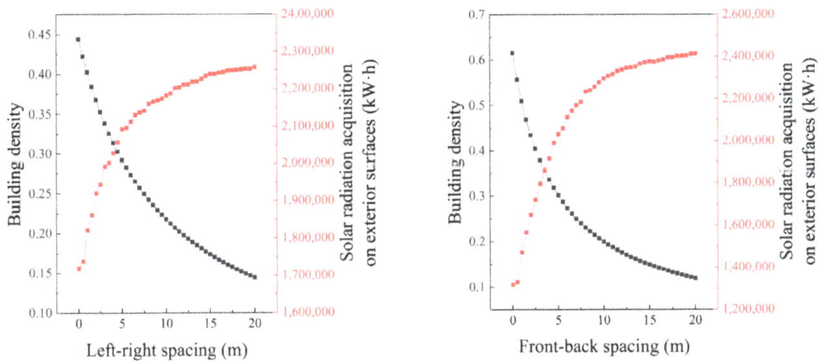

(**a**) Left–right arrangement (**b**) Front–back arrangement

Figure 14. Impact of spacing and BD on solar radiation acquisition for two-story buildings.

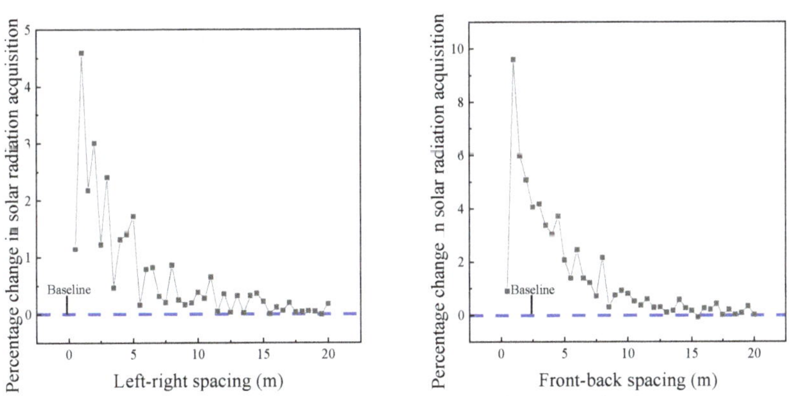

(**a**) Left–right arrangement (**b**) Front–back arrangement

Figure 15. Effect of spacing on percentage change in solar radiation acquisition for two-story buildings.

The thermal analysis (Figures 16–19) confirmed the dominant role of front-to-back spacing in energy regulation, with HEC stabilizing beyond 5 m for single-story buildings and 10 m for two-story buildings. The optimal spacing thresholds for energy efficiency—3–5 m for single-story and 5.5–10 m for two-story structures—balance radiation efficiency and land-use intensity, achieving an 18–22% reduction in HEC while maintaining 85–90% solar energy utilization.

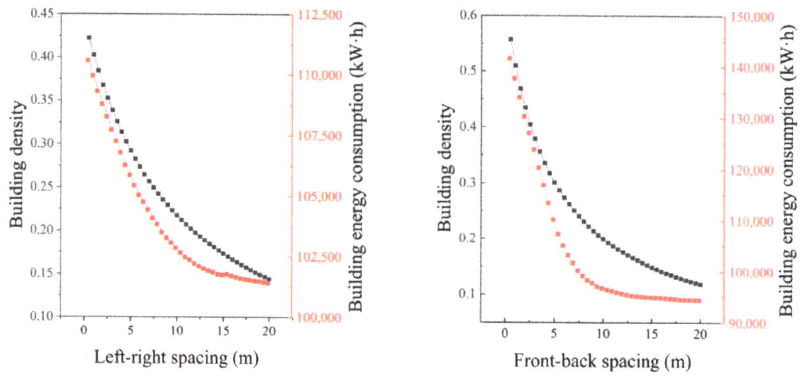

(**a**) Left–right arrangement (**b**) Front–back arrangement

Figure 16. Impact of building spacing on HEC of two-story buildings.

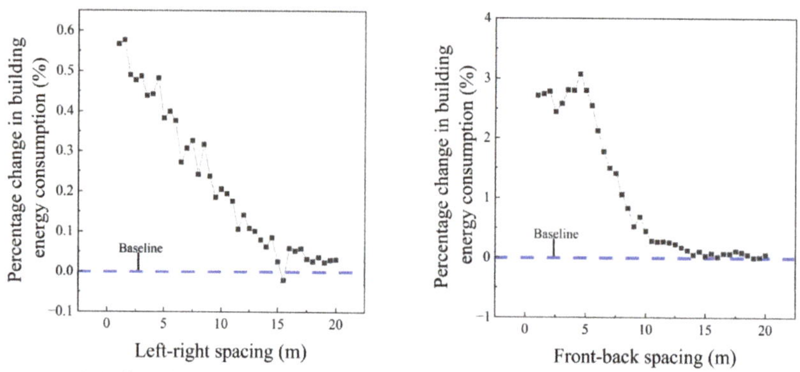

(**a**) Left–right arrangement (**b**) Front–back arrangement

Figure 17. Impact of building pacing on percentage change in building HEC of two-story buildings.

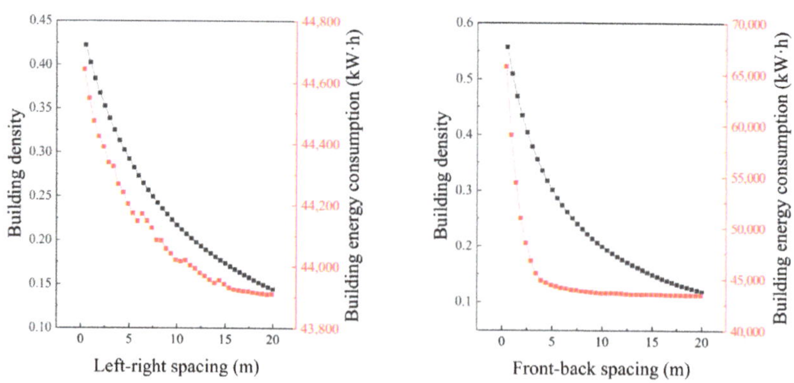

(**a**) Left–right arrangement (**b**) Front–back arrangement

Figure 18. Impact of building spacing on HEC of single-story buildings.

These findings establish a differentiated planning framework based on building height; low-rise structures adopt a compact east–west spacing layout, while multi-story configurations require expanded north–south spacing.

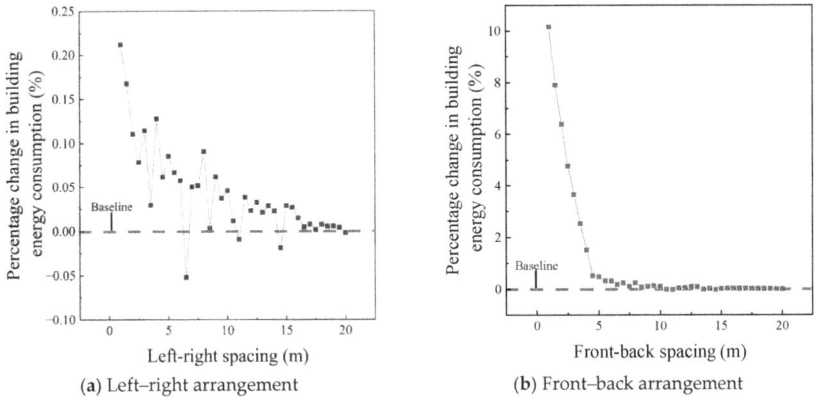

Figure 19. Impact of building pacing on percentage change in building HEC of single-story buildings.

4. Conclusions

This study took Qinghai as an example to systematically analyze the influence mechanisms of spatial structure on solar radiation acquisition and building HEC in high-altitude cold region resettlement communities.

Building spacing is a key factor regulating solar radiation acquisition and HEC; the impact of north–south spacing is significantly greater than that of east–west spacing, with improvements in solar radiation acquisition and HEC stabilizing beyond 5 m for single-story and 10 m for two-story buildings. Optimal spacing thresholds were derived from simulations as follows: for single-story buildings, 3–4 m north–south and 1–1.5 m east–west; for two-story buildings, 5–10 m north–south and 3–5 m east–west. Terraced layouts minimize exposed surfaces and can significantly reduce HEC.

Orientation optimization exhibits scale-dependent characteristics; individual buildings achieve peak solar gain at 10° west of south, while settlement clusters perform best with a due south orientation to reduce mutual shading. Recommended energy-efficient orientation sectors are from 50° east of south to 40° west of south for individual buildings, and from 60° east of south to 60° west of south for settlement planning, balancing radiation efficiency and spatial adaptability.

Courtyard depth induces nonlinear radiation responses; depths below 3 m form a high-sensitivity zone (radiation recovery rate between 55.7% and 79.6%), while depths beyond 4.8 m enter a stable zone (maintaining about 86.6% of the baseline). Thermal performance analysis further confirms 2.7 m as a critical depth threshold for minimizing heating losses.

Based on these findings, the study proposes an integrated "Orientation–Spacing–Depth" hierarchical design framework prioritizing south-facing façades, adapting spacing thresholds to building height, and modulating courtyard scale functionally. This climate-responsive blueprint demonstrates transfer potential to regions sharing Qinghai's high-altitude cold arid characteristics—specifically, those with comparable solar radiation regimes, heating-dominated energy profiles, and low-rise clustered morphologies.

5. Limitations and Prospects

This study intentionally focuses on single-variable mechanisms through idealized scenarios, thereby excluding multi-variable synergies, socioeconomic trade-offs, and dynamic complexities. Three primary technical constraints warrant attention in broader applications, as follows: radiation thresholds require latitude-specific adjustments; courtyard depth effects may reverse in high-humidity cold climates; and spacing optima assume homogeneous building heights. Critically, while our morphology–energy guidelines establish preventive design thresholds for early planning phases, their economic viability necessitates future region-specific techno-economic verification, a prerequisite for sustainable implementation. Subsequent research must develop integrated frameworks coupling environmental objectives with socioeconomic factors, enabling empirical validation across diverse high-altitude contexts to bridge theoretical insights and community-scaled practices.

Author Contributions: Conceptualization, B.L., W.S.; methodology, B.L. and Y.L.; software, B.L. and W.S.; validation, B.L. and J.S.; formal analysis, Y.L., W.S., and C.W.; investigation, B.L., Y.L., W.S., C.W., and J.S.; resources, Y.L., W.S., and C.W.; data curation, W.S., C.W., and J.S.; writing—original draft preparation, B.L., W.S.; writing—review and editing, B.L. and Y.L.; visualization, B.L.; supervision, Y.L., W.S., C.W., and J.S.; project administration, Y.L., W.S., and C.W.; funding acquisition, C.W. All authors have read and agreed to the published version of the manuscript.

Funding: This research was funded by Power China Northwest Engineering Corporation Limited through the key scientific research project [Number: XBY-KJ-2023-21].

Data Availability Statement: The original contributions presented in this study are included in the article. Further inquiries can be directed to the corresponding authors.

Acknowledgments: The authors acknowledge Power China Northwest Engineering Corporation Limited for their support of this research.

Conflicts of Interest: Author Wei Song and Chuanming Wang were employed by the company Power China Northwest Engineering Corporation Limited. The remaining authors declare that the research was conducted in the absence of any commercial or financial relationships that could be construed as a potential conflict of interest.

References

1. Ren, J.; Wang, Y.; Liu, Q.; Liu, Y. Numerical Study of Three Ventilation Strategies in a prefabricated COVID-19 inpatient ward. *Build. Environ.* **2021**, *188*, 107467. [CrossRef]
2. Liu, B.; Liu, Y.; Cho, S.; Chow, D.H.C. Urban Morphology Indicators and Solar Radiation Acquisition: 2011–2022 Review. *Renew. Sustain. Energy Rev.* **2024**, *199*, 114548. [CrossRef]
3. Xiao, Q.; Liu, Y.; Cho, S.; He, Y. Development of a Cross-Latitude Applicable Predictive Model for Solar Radiation Acquisition Potential of Urban Residential Blocks. *Energy Build.* **2024**, *320*, 114594. [CrossRef]
4. Manni, M.; Nocente, A.; Kong, G.; Skeie, K.; Fan, H.; Lobaccaro, G. Solar Energy Digitalization at High Latitudes: A Model Chain Combining Solar Irradiation Models, a LiDAR Scanner, and High-Detail 3D Building Model. *Front. Energy Res.* **2022**, *10*, 1082092. [CrossRef]
5. Freitas, S.; Catita, C.; Redweik, P.; Brito, M.C. Modelling Solar Potential in the Urban Environment: State-of-the-Art Review. *Renew. Sustain. Energy Rev.* **2015**, *41*, 915–931. [CrossRef]
6. Gan, V.J.L.; Wong, H.K.; Tse, K.T.; Cheng, J.C.P.; Lo, I.M.C.; Chan, C.M. Simulation-Based Evolutionary Optimization for Energy-Efficient Layout Plan Design of High-Rise Residential Buildings. *J. Clean. Prod.* **2019**, *231*, 1375–1388. [CrossRef]
7. He, P.; Xue, J.; Shen, G.Q.; Ni, M.; Wang, S.; Wang, H.; Huang, L. The Impact of Neighborhood Layout Heterogeneity on Carbon Emissions in High-Density Urban Areas: A Case Study of New Development Areas in Hong Kong. *Energy Build.* **2023**, *287*, 113002. [CrossRef]
8. Dino, I.G.; Üçoluk, G. Multiobjective Design Optimization of Building Space Layout, Energy, and Daylighting Performance. *J. Comput. Civ. Eng.* **2017**, *31*, 04017025. [CrossRef]
9. Hachem-Vermette, C.; Singh, K. Mixed-Use Neighborhoods Layout Patterns: Impact on Solar Access and Resilience. *Sustain. Cities Soc.* **2019**, *51*, 101771. [CrossRef]
10. Asfour, O.S.; Alshawaf, E.S. Effect of Housing Density on Energy Efficiency of Buildings Located in Hot Climates. *Energy Build.* **2015**, *91*, 131–138. [CrossRef]
11. Košir, M.; Capeluto, I.G.; Krainer, A.; Kristl, Ž. Solar Potential in Existing Urban Layouts—Critical Overview of the Existing Building Stock in Slovenian Context. *Energy Policy* **2014**, *69*, 443–456. [CrossRef]
12. Deng, Q.; Wang, G.; Wang, Y.; Zhou, H.; Ma, L. A Quantitative Analysis of the Impact of Residential Cluster Layout on Building Heating Energy Consumption in Cold IIB Regions of China. *Energy Build.* **2021**, *253*, 111515. [CrossRef]

13. Xu, S.; Li, G.; Zhang, H.; Xie, M.; Mendis, T.; Du, H. Effect of Block Morphology on Building Energy Consumption of Office Blocks: A Case of Wuhan, China. *Buildings* **2023**, *13*, 768. [CrossRef]
14. Ahmed, W.; Asif, M.; Alrashed, F. Application of Building Performance Simulation to Design Energy-Efficient Homes: Case Study from Saudi Arabia. *Sustainability* **2019**, *11*, 6048. [CrossRef]
15. Yan, J.; Zhang, H.; Li, Y.; Huang, X.; Jin, S.; Jia, X.; Ke, Z.; Yu, H. Study on the Influence of the Energy Intensity of Residential District Layout on Neighborhood Buildings. *Sustainability* **2023**, *15*, 15307. [CrossRef]
16. Narimani Abar, S.; Schulwitz, M.; Faulstich, M. The Impact of Urban Form and Density on Residential Energy Use: A Systematic Review. *Sustainability* **2023**, *15*, 15685. [CrossRef]
17. De Rubeis, T.; Giacchetti, L.; Paoletti, D.; Ambrosini, D. Building Energy Performance Analysis at Urban Scale: A Supporting Tool for Energy Strategies and Urban Building Energy Rating Identification. *Sustain. Cities Soc.* **2021**, *74*, 103220. [CrossRef]
18. Du, T.; Turrin, M.; Jansen, S.; Van den Dobbelsteen, A.; De Luca, F. Relationship Analysis and Optimisation of Space Layout to Improve the Energy Performance of Office Buildings. *Energies* **2022**, *15*, 1268. [CrossRef]
19. Zhu, S.; Ma, C.; Wu, Z.; Huang, Y.; Liu, X. Exploring the Impact of Urban Morphology on Building Energy Consumption and Outdoor Comfort: A Comparative Study in Hot-Humid Climates. *Buildings* **2024**, *14*, 1381. [CrossRef]
20. Hadavi, M.; Pasdarshahri, H. Investigating Effects of Urban Configuration and Density on Urban Climate and Building Systems Energy Consumption. *J. Build. Eng.* **2021**, *44*, 102710. [CrossRef]
21. Wu, P.; Liu, Y. Impact of Urban Form at the Block Scale on Renewable Energy Application and Building Energy Efficiency. *Sustainability* **2023**, *15*, 11062. [CrossRef]
22. Hong, B.; Lin, B. Numerical Studies of the Outdoor Wind Environment and Thermal Comfort at Pedestrian Level in Housing Blocks with Different Building Layout Patterns and Trees Arrangement. *Renew. Energy* **2015**, *73*, 18–27. [CrossRef]
23. Yan, J.; Zhang, H.; Liu, X.; Ning, L.; Hien, W.N. The Impact of Residential Cluster Layout on Building Energy Consumption and Carbon Emissions in Regions with Hot Summers and Cold Winters in China. *Sustainability* **2023**, *15*, 11915. [CrossRef]
24. Lu, M.; Zhang, Y.; Xing, J.; Ma, W. Assessing the Solar Radiation Quantity of High-Rise Residential Areas in Typical Layout Patterns: A Case in North-East China. *Buildings* **2018**, *8*, 148. [CrossRef]
25. Singh, K.; Hachem-Vermette, C.; D'Almeida, R. Solar Neighborhoods: The Impact of Urban Layout on a Large-Scale Solar Strategies Application. *Sci. Rep.* **2023**, *13*, 18843. [CrossRef]
26. Sosa, M.B.; Correa, E.N.; Cantón, M.A. Neighborhood Designs for Low-Density Social Housing Energy Efficiency: Case Study of an Arid City in Argentina. *Energy Build.* **2018**, *168*, 137–146. [CrossRef]
27. Huo, K.; Qin, R.; Zhao, J.; Ma, X. Long-term tracking of urban structure and analysis of its impact on urban heat stress: A case study of Xi'an, China. *Ecological Indicators* **2025**, *174*, 113418. [CrossRef]
28. *GB 55015-2021*; Ministry of Housing and Urban-Rural Development. General Code for Energy Efficiency and Renewable Energy Application in Buildings. China Architecture & Building Press: Beijing, China, 2021.
29. *GB/T 51350-2019*; Ministry of Housing and Urban-Rural Development. Technical Standard for Nearly Zero Energy Buildings. China Architecture & Building Press: Beijing, China, 2019.

buildings

MDPI

Article

Correlation Between the Insolation Shadow Ratio and Thermal Comfort of Urban Outdoor Spaces in Residential Areas in Xi'an

Jie Song [1,2], Yu Liu [2,3,*], David Hou Chi Chow [4], Bo Liu [5] and Seigen Cho [6]

1 School of Mechanical Engineering, Northwestern Polytechnical University, Xi'an 710072, China; songjay@mail.nwpu.edu.cn
2 Sustainable Building and Environmental Research Institute, Northwestern Polytechnical University, Xi'an 710129, China
3 School of Mechanics, Civil Engineering and Architecture, Northwestern Polytechnical University, Xi'an 710129, China
4 School of Architecture, University of Liverpool, Liverpool L69 7ZX, UK
5 School of Architecture, Xi'an University of Architecture and Technology, Xi'an 710055, China
6 Institute of Urban Innovation, Yokohama National University, Yokohama 240-8501, Japan
* Correspondence: liuyu@nwpu.edu.cn

Abstract: Solar exposure and shading critically influence outdoor thermal comfort in residential areas, yet quantitative links between spatial morphology and microclimate remain insufficiently explored in cold-region cities. This study proposes a novel morphological indicator, the Insolation Shadow Ratio (ISR), to quantify sunlight–shade dynamics and investigates its correlation with outdoor thermal comfort (UTCI) in Xi'an, China. Combining field observations, microclimate simulations, and statistical analysis, we quantified ISR and UTCI across three representative outdoor spaces in a residential area. Photographic analysis and spatial parameterization were employed to calculate hourly ISR values. Significant correlations were observed between ISR and UTCI values. The measured data showed the strongest correlation at summer solstice at site C (Spearman's r = 0.883, $p < 0.01$). GAM analysis of seasonal peak correlation data revealed that an optimal UTCI comfort range of 9 °C to 26 °C, corresponding to ISR thresholds of 0.0202–0.8384, achieved the highest autumn correlation at site C (r = 0.686, $p < 0.01$), while effectively balancing shade cooling effects and solar accessibility. The ISR framework provides a quantifiable tool for designers to optimize outdoor thermal environments and, when enhanced by parametric modeling tools, enables them to proactively optimize thermal performance during early-stage residential planning, offering a data-driven pathway for climate-resilient outdoor space design.

Keywords: Insolation Shadow Ratio; thermal comfort; UTCI; residential area; outdoor space

1. Introduction

The rapid densification of urban residential areas in cold-region cities like Xi'an has significantly altered solar exposure patterns, exacerbating seasonal thermal stress in outdoor spaces [1]. While sunlight enhances winter comfort, excessive solar radiation in summer intensifies urban heat island effects, reducing space usability. Effective management of sunlight–shadow dynamics is thus critical for climate-responsive design. Some studies highlighted significant correlations between solar exposure [2], shadow coverage [3], and thermal comfort in urban spaces [4,5], emphasizing the importance of balancing sunlight and shading management [6–8]. Compagnon demonstrated the significant impact of solar

https://doi.org/10.3390/buildings15121995

and shadow movement on outdoor thermal comfort through software simulations [9,10]. Acero et al. investigated seasonal variations in urban sunlight and shadows, revealing their profound impact on thermal comfort across different seasons [11]. Advanced modeling techniques, such as Structural Equation Modeling (SEM), have quantified the impacts of sunlight and shading on thermal comfort, offering valuable insights for urban planning [12]. Numerous studies have explored strategies to improve urban outdoor thermal conditions [13]. Modification of solar irradiation conditions through shading, with higher levels of shade in summer, improve thermal comfort [7,14,15]. Thermal comfort indices such as the Physiological Equivalent Temperature (PET) and the Universal Thermal Climate Index (UTCI) have been widely used to evaluate outdoor thermal environments [16]. These indices consider factors like solar exposure and shading to provide comprehensive assessments of thermal conditions [17–19].

Urban morphology governs the distribution of solar access and shade, both of which critically shape outdoor thermal comfort. Current studies on morphological indicators and thermal comfort have widely adopted the Sky View Factor (SVF) and Floor Area Ratio (FAR) as core metrics to quantify the relationship between urban morphology and thermal environments [20]. SVF, calculated as the proportion of visible sky at a given point, reflects the openness of urban canyons and has demonstrated significant impacts on outdoor thermal comfort. For instance, higher SVF values correlate with elevated UTCI in unshaded areas due to increased solar exposure [21,22]. FAR, defined as the ratio of total building volume to site area, controls urban density and indirectly influences shadow coverage. While increasing FAR enhances summer shadow coverage [23], it may exacerbate insufficient daylight penetration in winter, thereby compromising seasonal thermal balance [24,25]. Morphological indicators like the SVF and FAR optimize thermal comfort by controlling solar access and enhancing airflow [16]. Together, these interdependent factors—form, shade, and microclimate—underscore the need for a dynamic, geometry-driven indicator that links spatial configuration directly to thermal comfort performance.

Considering other morphological indicators and thermal comfort studies, Song et al. [15] found that building orientation and height play critical roles in optimizing solar exposure and shadow projection. Yi et al. [23] explored the relationship between urban density and sunlight/shadow patterns, noting that denser areas typically reduce sunlight exposure while increasing shading, thereby influencing outdoor comfort.

However, current urban planning practices often rely on generic metrics like the SVF and FAR, which inadequately capture time-varying solar–shadow interactions at human scales [18,21,26]. Prior studies have demonstrated strong links among solar exposure, shadow coverage, and thermal comfort—highlighting the role of factors such as vegetation density [2], building aspect ratio [15], and material albedo [27]—yet common morphological metrics like the SVF and FAR remain static and cannot capture diurnal and seasonal solar trajectories. This gap motivates the development of ISR, a time-sensitive indicator that dynamically integrates solar azimuth and altitude to quantify hourly sunlit fractions, providing a continuous, geometry-driven proxy for outdoor thermal comfort. Furthermore, while parametric tools like ENVI-met enable detailed microclimate simulations, their complexity limits integration into early-stage design workflows. Crucially, no study has proposed a time-sensitive morphological indicator that balances solar accessibility and shading requirements across seasons while remaining operable for architects and planners. To address these limitations, this study proposes the Insolation Shadow Ratio (ISR), defined as the hourly ratio of sunlit area to total site area. Unlike the SVF or FAR, ISR dynamically quantifies solar–shadow interplay by explicitly incorporating hourly variations in solar position, including azimuth and altitude, across different seasons. This dynamic consideration enables ISR to capture temporal patterns of solar exposure and shading that are overlooked by static indicators such as the SVF. Conceptually, while SVF is a static ratio representing the proportion of visible sky dome regardless of time, ISR is defined as the ratio of sunlit area to total site area at a specific hour, which allows for continuous assessment of solar access and shading dynamics throughout the diurnal and seasonal cycles, thereby enhancing its applicability in thermal comfort-oriented urban design. The dynamic trajectory of solar position across seasonal and diurnal cycles (with hourly variations) serves to quantify the spatiotemporal patterns of solar exposure and shadowing at multiple temporal scales (e.g., summer solstice, winter solstice, and equinoxes), thereby enabling precise characterization of the dynamic impacts of architectural morphology on outdoor thermal environments. Ground-level shadow patterns are shaped by building morphology, with radiative heat exchange mechanisms demonstrating a measurable correlation between these patterns and UTCI. This evidence establishes the critical role of shadow dynamics in urban thermal regulation.

The influence of building shadows on urban microclimates has been widely explored in recent years. Lindberg et al. [28] demonstrated that shading can significantly reduce mean radiant temperature (T_MRT) in residential areas, thereby improving outdoor thermal comfort. Prior research in diverse climatic regions has employed simulation-based models such as SOLWEIG and ENVI-met to assess the thermal effects of solar access and shadow distribution. For instance, studies in the Phoenix–Tempe metropolitan area used SOLWEIG to generate high-resolution shadow and T_MRT maps to support summer heat mitigation and urban cooling strategies. In China, the Building Shadow Generation Ratio (BSGR) was proposed as a dynamic alternative to the Sky View Factor (SVF), offering improved evaluation of microscale thermal conditions in residential environments [16].

However, most existing studies are constrained by two key limitations. First, they typically focus on single-season optimization—primarily summer—without addressing the seasonal contradictions in temperate regions where summer shading needs to be balanced against winter solar access. Second, morphological–climatic interactions are often modeled using static indicators or rule-based parameters, lacking a temporally explicit, quantitative framework for analyzing shadow dynamics across seasons.

The present study addresses these gaps through three novel contributions. First, by selecting Xi'an—a representative temperate monsoon city with four distinct seasons—the study investigates solar–shadow interactions across solstices and equinoxes, enabling a seasonally balanced analysis of thermal comfort. Second, the proposed Insolation Shadow Ratio (ISR) offers a time-sensitive morphological indicator that captures diurnal and seasonal variations in solar altitude and azimuth. Its correlation with thermal comfort (UTCI) is quantified using both field measurements and generalized additive modeling (GAM), providing improved temporal resolution over static metrics such as SVF and BSGR. Third, the derived thresholds offer a quantifiable link between spatial form and environmental performance, supporting climate-responsive urban design beyond the capabilities of existing models.

This study aims to establish ISR as a novel morphological indicator for solar–shadow management and quantify correlations between ISR and UTCI across diverse residential layouts. Therefore, the research contents of this study corresponding to the above-raised limitations are: (1) serving as a design-friendly metric for real-time shadow optimization using parametric tools (Grasshopper + Ladybug), (2) analyzing the relationship between ISR and UTCI to validate ISR as an indicator for managing outdoor thermal comfort, and (3) providing ISR thresholds for a climate-specific area. By resolving the above research questions, ISR advances practical design workflows.

2. Methodology

2.1. Research Framework

Focusing on three high-activity hotspots (western, north-central, and southeastern zones) on a mixed-use residential campus (prioritizing residential over academic areas), we defined "high-activity" areas as those exhibiting pedestrian flows above 30 persons/hour during peak periods, corroborated by semi-structured interviews with residents. At each hotspot, we then integrated field measurements, parametric modeling (Grasshopper + Ladybug), and statistical analysis to assess dynamic solar–shadow interactions and outdoor thermal comfort. Hourly ISR values were derived from spatial geometry parameterization and time-series solar radiation simulations, while UTCI was calculated using localized meteorological data. The research framework (Figure 1) consists of several steps: ① Field Observations and Interviews: Site surveys and resident interviews were conducted to identify frequently used outdoor spaces of residential area. Key environmental features, such as building enclosure form, shading patterns, and vegetation, were documented. ② Data Collection: Climatic variables including air temperature, relative humidity, wind speed, and mean radiant temperature were measured. Meanwhile, photographic analysis and simulation tools, such as Ladybug and Honeybee in Grasshopper, were used to calculate solar exposure and shading patterns. ③ Calculation of Insolation Shadow Ratio (ISR): Calculation of the ISR was divided into two parts: the measured ISR was calculated based on the projected area of insolation shadows of the actual site, and the simulated ISR was obtained based on the established settlement models. ④ Access to ISR and UTCI Simulated Data: Calculation of ISR and UTCI simulated data by Ladybug and Honeybee. ⑤ Correlation Analysis: Statistical correlation analysis was performed to evaluate the relationship between ISR and UTCI. A combination of measured and simulated data was used to ensure accuracy and reliability. ⑥ Find the ISR Threshold: The seasons with high correlation are filtered out, and the comfort intervals of ISR and UTCI are fitted by the GAM model to obtain the corresponding ISR thresholds.

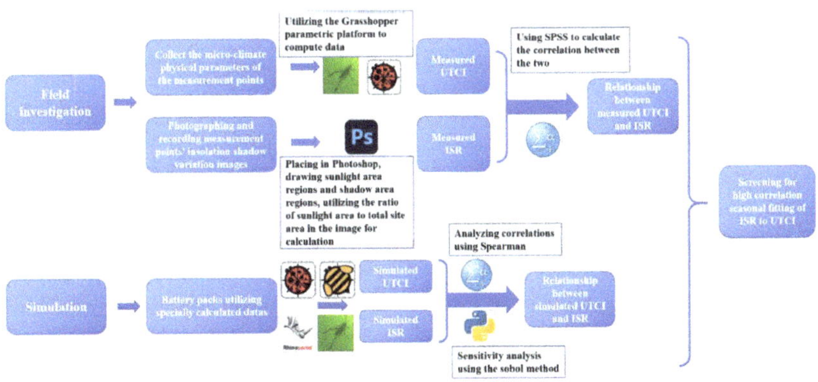

Figure 1. Research framework diagram.

2.2. Sites

Preliminary investigations identified three key areas within the neighborhood—the western region, the north-central region, and the southeastern region—as hotspots of resident activity. These areas were selected as representative observation sites. The neighborhood comprises both residential and teaching areas, and the study focused exclusively on the residential areas (Figure 2a,b).

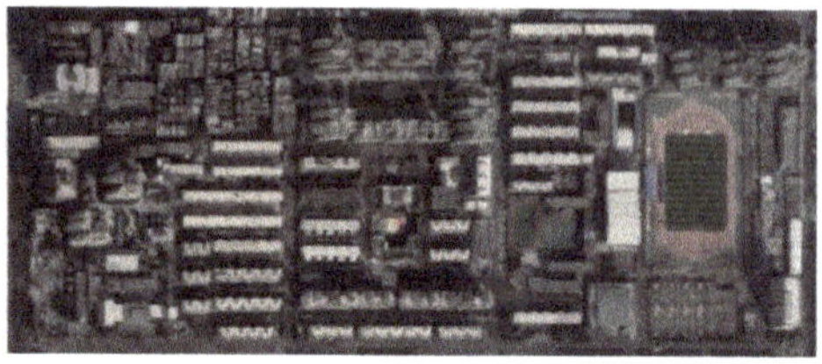

(a)

(b)

Figure 2. (**a**) Photo of the survey area (source: Baidu web map). (**b**) General plan of the survey area (source: Authors' record of residential area).

Site A is situated in the western part of the residential neighborhood and occupies a rectangular area approximately 17 m in width and 72 m in depth, covering a total area of about 1224 m^2. The site is surrounded by buildings five to seven stories high on its eastern and western sides, with scattered vegetation distributed across the area.

Site B is situated in the south-central part of the neighborhood and covers a rectangular area approximately 26 m in width and 40 m in depth, with a total area of about 1040 m^2. A 12-story residential building is located to the south of the site, with residential and office buildings. To the east, it is adjacent to a school campus. The site is surrounded by a mix mid- to high-rise buildings, public structures, and landscaped vegetation. Additionally, it is equipped with seating areas for pedestrians to rest.

Site C is situated within a street garden in the residential neighborhood. It is rectangular, measuring approximately 55 m in width and 32 m in depth, covering a total area of about 1760 square meters. To the west, it is bordered by six-story residential buildings, while the north side features an 18-story high-rise residential building. The east and south sides are adjacent to public buildings. The site is surrounded by trees and dense vegetation. Nearby, there is an affiliated kindergarten, contributing to an active pedestrian zone primarily frequented by elderly individuals engaging in exercise and parents accompanying their children.

2.3. Field Investigation

2.3.1. Objectives

The field investigation targeted two primary aspects. The first focused on outdoor climatic factors influencing residents' activities, such as air temperature, relative humidity, wind speed, solar radiation, and variations in sunlight and shadow. The aim was to understand how changes in these factors affect thermal comfort in real-life outdoor residential areas. The second aspect examined the surrounding building forms and their basic configurations to analyze how they shape outdoor environments.

2.3.2. Content and Methods

Using an existing residential neighborhood in Xi'an as a case, the study explored changes in sunlight and shadows areas in outdoor spaces under varying building configurations and how these changes impact thermal comfort of the space. Field measurements were conducted on key seasonal days—the winter solstice, the spring equinox, and the summer solstice. The winter solstice and the summer solstice represent extremes in solar radiation: minimal sunlight during the winter solstice and maximum sunlight during the summer solstice. These data provide insights into the influence of sunlight variations on outdoor spaces and environments. Due to the COVID-19 pandemic restrictions in 2022, autumn equinox measurements were unavailable. Instead, the spring equinox was included to bridge the seasonal gap between the solstices, offering a balanced perspective on seasonal sunlight conditions. The nearly 12 h daylight period during the spring equinox, similar to the autumn equinox, provided complementary data for analyzing ISR's influence on outdoor thermal comfort across seasons.

Observations were conducted between sunrise and sunset (8:00 a.m. to 6:00 p.m.), capturing sunlight–shadow patterns at two-hour intervals. Images documenting shadow distribution and residents' activities were recorded, alongside climatic variables such as air temperature, Wet Bulb Globe Temperature (WBGT), relative humidity, and wind velocity, at one-minute intervals. Behavioral observations complemented environmental measurements to provide insights into how residents interacted with the environment.

2.4. Selection of Outdoor Thermal Comfort Indicators

Since the mid-20th century, outdoor thermal comfort evaluation methods have evolved significantly—from simplified heat balance models to more physiologically accurate indices. Early research identified key environmental variables such as air temperature, relative humidity, wind speed, and mean radiant temperature, alongside physiological factors like clothing insulation and metabolic rate, as fundamental determinants of human thermal comfort [29–31]. When exploring the relationship between outdoor thermal comfort and solar–shadow conditions, it is essential to select appropriate indicators for evaluation [32,33]. These foundational insights led to the development of steady-state indices, including the Physiological Equivalent Temperature (PET) [34,35], Standard Effective Temperature (SET) [36–38], and Universal Thermal Climate Index (UTCI) [39–42].

More recently, the Universal Thermal Climate Index (UTCI) was introduced, based on a dynamic multi-node thermoregulation model that accounts for human physiological responses under fluctuating outdoor conditions [39,40]. Compared to steady-state indices, UTCI provides improved accuracy in capturing the transient effects of solar radiation and wind, making it particularly suitable for outdoor environments [40,43,44].

In China, researchers such as Jin Hong [45], Nie Ting [46], and Wu Jinkui [47] have studied the applicability of the UTCI across different regions. Wu Zhifeng [48] demonstrated its relevance in analyzing thermal conditions under four greening retrofit scenarios in residential areas during typical summers. Wereski Sylwester [42] analyzed the spatial and temporal variations of the UTCI during 21st century winters in Poland, indicating that the UTCI can also be used to assess large-scale thermal comfort. Park Sookuk [49] argued that the UTCI is a widely accepted and easily determined thermal index, making it a valuable tool for analyzing outdoor thermal effects on humans in urban and landscape planning and design.

In summary, the UTCI is not only suitable for evaluating outdoor thermal comfort, but also capable of revealing human thermal sensations under varying environmental conditions [50]. Given its dynamic modeling framework and strong empirical support, UTCI is adopted in this study as the core indicator for assessing outdoor thermal comfort. It integrates key climatic variables—air temperature, humidity, wind speed, and mean radiant temperature—to provide a comprehensive representation of human thermal sensation. In this research, UTCI was computed through parametric simulations using Grasshopper, enabling spatially and temporally refined analysis of thermal comfort conditions in relation to the ISR [25].

2.4.1. Method for Calculating the UTCI Based on Field Measurements

To obtain the Universal Thermal Climate Index (UTCI) for the measured residential area, the "LB UTCI Comfort" component in Ladybug was used to calculate the UTCI. The simulation assumed a standing posture (1.2 met) docs.ladybug.tools (accessed on 21 June 2024), with seasonal clothing insulation set to 0.5 clo for summer and 1.0 clo for winter, based on typical ensemble values recommended by ASHRAE 55. The input data include the measured values for air temperature, mean radiant temperature (MRT), relative humidity, and wind velocity at various measurement points to calculate the UTCI for the residential area (as shown in Figure 3).

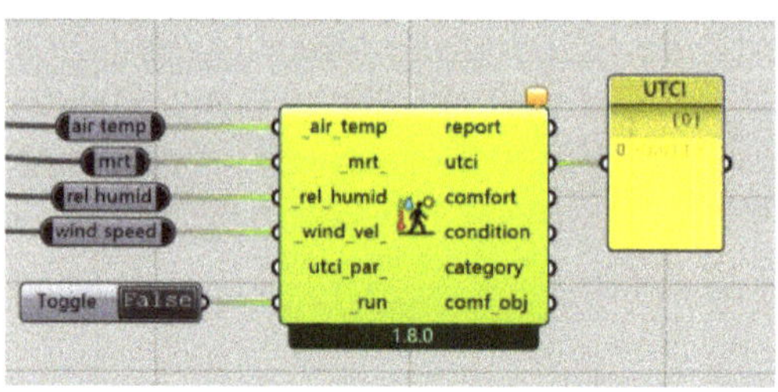

Figure 3. Ladybug UTCI computing battery.

2.4.2. Method for Calculating the UTCI Based on Simulation

Based on the building types and enclosure characteristics of site A, site B, and site C in the measured residential area, 3D models of the three sites were created. To explore the relationship between UTCI data and the ISR under ideal model conditions, Ladybug components were used to acquire and analyze site data. The constructed building models of the sites are input as the analysis objects. The corresponding weather files for the simulation's climate zone are selected, and the weather data is input into the "LB Solar Outdoor MRT" component to calculate the mean radiant temperature (MRT). The obtained MRT, along with other weather data, is then input into the component for calculating the UTCI. Finally, through simulation analysis of the building models, the average, minimum, and maximum UTCI values for the study period are obtained on an hourly basis. Establishing a 5 × 5 m grid based on the site area facilitates picking up variation of the UTCI in the area, so that the maximum, minimum, and average values of UTCI for the site can be obtained through battery calculations.

2.5. Concept and Calculation Method of the ISR Index

This study introduces a new morphological index, the Insolation Shadow Ratio (ISR), to quantify the variation between sunlight and shadow in outdoor residential areas. It is defined as the ratio of the sunlit area to the total area of the site. In this study, the shadow area is primarily composed of the shadows cast by buildings onto the ground area of the outdoor site under solar radiation. As shown in Figure 4, "As" represents the shadow area, "Ai" represents the sunlit area, and the green squares represent the building. The calculation formula is as follows:

$$ISR = \frac{A_i}{(A_i + A_s)}$$

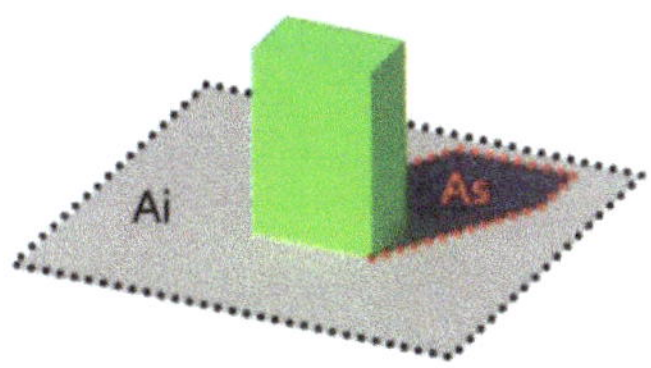

Figure 4. Schematic diagram of the ISR (illustration by the author).

By explicitly aggregating overlapping shadows into As, ISR remains robust in complex urban settings where multiple structures cast intersecting shadow patterns. Figure 5 provides a schematic comparison of ISR calculations for single-building and multi-building configurations, clearly illustrating how group shadow masking is accounted for.

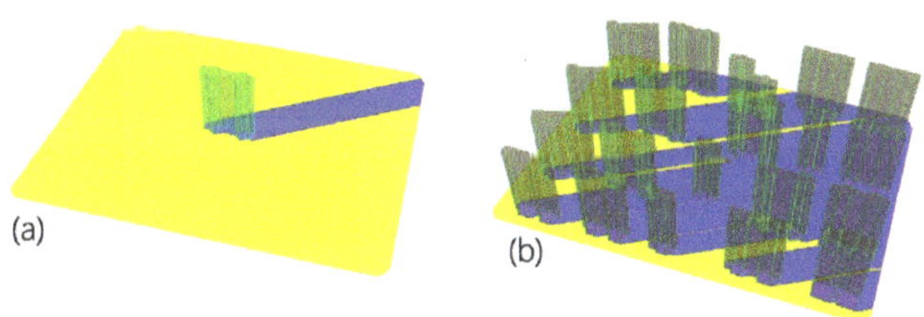

Figure 5. (**a**) ISR calculation for a single building. (**b**) ISR calculation for a group of buildings (illustration by the author).

2.5.1. Method for Obtaining ISR Values Through Field Measurements

The method for obtaining ISR values through field measurements involved observing variation in sunlight and shadow on the site. This was done by using panoramic photographs and image recognition techniques to record changes in sunlight and shadow. The contrast images of sunlight and shadow were extracted according to time periods, allowing calculation of the ratio of the sunlit area to the total site area. The typical solar terms were selected as the representative days of the seasons. Therefore, field data were collected on 22 December 2022 (the winter solstice), 21 March 2023 (the spring equinox), and 22 June 2023 (the summer solstice).

2.5.2. Method for Obtaining ISR Values Through Simulation

Grasshopper is a tool that integrates visual programming and building environmental performance analysis. It is typically used in conjunction with Rhino 3D to create parametric designs and perform various computational tasks. By utilizing built-in components in

Grasshopper, such as solar trajectory and sunlight angle functions, the ratio of ground area receiving sunlight to the total site area can be calculated.

The geometric shapes to be analyzed, such as buildings, sites, or other structures requiring sunlight and shadow ratio simulations, were first input into Grasshopper. Subsequently, Grasshopper was configured, and a new definition for the ISR was created. The parameters necessary for the simulation, including location data (latitude, longitude, date, and time), solar attributes (solar position and radiation data), and other relevant information (e.g., terrain, site, and buildings), were then defined. Using the solar analysis components of the Honeybee tool, calculations were performed to determine sunlight exposure, shadow ratios, and other solar-related data, such as solar radiation and shadow analysis. Finally, visualizations of the ISR were generated, presenting this information graphically through Grasshopper components.

2.6. Validation of the Relationship Between Outdoor Thermal Comfort and the ISR

This study aims to conduct a quantitative accuracy assessment between field-measured data and simulated outputs. Correlation and sensitivity analyses between ISR and UTCI need to be performed. However, given the inherent environmental dynamics (e.g., spatiotemporal heterogeneity of meteorological variables) and persistent anthropogenic disturbances (e.g., land-use alterations and infrastructure interventions) observed in field conditions, coupled with data availability constraints, the sensitivity analysis is confined to evaluating seasonal variations within the modeling framework. This targeted approach allows systematic exploration of model robustness under controlled temporal parameters while acknowledging limitations in empirical data acquisition. Therefore, the sensitivity analysis was performed only for the simulated data.

2.6.1. Spearman Correlation

The correlation analysis framework adopts a hypothesis-agnostic approach to statistically identify key covariates exhibiting coordinated variation patterns. By systematically quantifying bivariate associations (Spearman's $|r| > 0.5$, $p < 0.01$), this methodology simultaneously addresses two operational objectives: (a) revealing latent variable co-dependencies through dimension-reduced projection, and (b) establishing empirically grounded variable screening criteria for subsequent mechanistic modeling [51–53].

The aim of this study is to explore whether there is a correlation between outdoor thermal comfort and ISR in residential areas of Xi'an. To validate the correlation and determine the strength of the relationship, normality tests and correlation analysis were conducted on both the field measurement data and simulation data using SPSS 26. For the correlation between the variables, Spearman's method was selected.

2.6.2. Sobol Sensitivity Analysis

The sensitivity analysis protocol was strategically designed to elucidate causal mechanisms and facilitate model optimization by systematically addressing a pivotal question: how can parametric adjustments enhance system performance [54–56]? By leveraging the Sobol variance decomposition method—a globally convergent sensitivity analysis technique—this study quantitatively evaluates both individual parameter effects (first-order indices, Si) and higher-order interaction impacts (total-order indices, ST) through variance-based importance metrics. Implemented via the SALib Python (3.12) package, this approach overcomes the limitations of local sensitivity analysis (LSA) through quasi-Monte Carlo sampling (Saltelli sequence, N = 1000 iterations), with parameter importance rankings validated via convergence diagnostics (Anderson-Darling test, $p > 0.1$) [57]. From the point of view of the amount of data, as well as the questions explored, this study conducted a sensitivity analysis only for the simulated data.

3. Results

3.1. Field Measurement Results

3.1.1. Measured Results of the Insolation Shadow Ratio for Different Types of Spaces

The data obtained from field measurements were calculated, and the ISR data are detailed in Table 1. Panoramic photographs documenting changes in sunshine and shadow at the site were collected during the field survey. Examples of the field recordings for the insolation shadow variation at site A are shown in Appendix A Tables A1–A3.

Table 1. Changes in the insolation shadow ratio during different seasons at site A.

Time Period	Winter Solstice—ISR	Spring Equinox—ISR	Summer Solstice—ISR
8:00–10:00	9.7%	46.33%	12.66%
10:00–12:00	40.66%	87.16%	30.45%
12:00–14:00	43.97%	93.53%	23.94%
14:00–16:00	1.06%	67.67%	67.21%
16:00–18:00	0%	43.59%	41.08%

3.1.2. Measured UTCI Results for Different Types of Spaces

(1) Comparison of measured UTCI data

During the winter solstice, the average UTCI for site A was 9.95 °C, significantly higher than site B's 7.30 °C and site C's 6.71 °C. The higher UTCI value for site A indicates that, although there was more sunlight in winter, the thermal comfort level remained relatively low, likely due to lower temperatures. Site C had the lowest UTCI, reflecting the limited sunlight conditions and thermal environment in winter.

During the summer solstice, site C had the highest average UTCI, at 39.71 °C, indicating the worst thermal comfort, possibly due to stronger solar radiation or insufficient shading. Site A followed, with an average UTCI of 38.89 °C, while site B had the lowest UTCI value of 37.42 °C, suggesting that site B may have better shading or architectural forms that contribute to microclimate regulation.

During the spring equinox, the UTCI values for site A (20.82 °C) and site C (20.72 °C) were similar, indicating comparable thermal comfort in spring. Site B's UTCI value was slightly lower, at 19.47 °C, possibly due to site layout or shading coverage.

(2) Analysis of UTCI Differences Between Sunlit and Shaded Areas

The measured data were visualized through graphical representation, as shown in the figures below. These figures illustrate the variations in sunlit and shaded areas for different sites across typical solar terms, including the summer solstice, the winter solstice, and the spring equinox. Specifically, they depicted the sunlit and shaded distributions for site A during the summer solstice (SSA-Sunlit, SSA-Shaded), the winter solstice (WSA-Sunlit, WSA-Shaded), and the spring equinox (SEA-Sunlit, SEA-Shaded); for site B under the same solar terms (SSB-Sunlit, SSB-Shaded, WSB-Sunlit, WSB-Shaded, SEB-Sunlit, SEB-Shaded); and for site C under identical conditions (SSC-Sunlit, SSC-Shaded, WSC-Sunlit, WSC-Shaded, SEC-Sunlit, SEC-Shaded). These visualizations clearly depicted seasonal variations in solar exposure across different sites and served as foundational data for further analysis of the correlation between ISR and outdoor thermal comfort.

During the winter solstice, the UTCI for the sunlit area of site A was significantly higher than that for the shaded area, with the positive effect of sunlight on thermal comfort being particularly evident. During the summer solstice, the UTCI for the sunlit area was also significantly higher than for the shaded area, indicating that sunlight exposure has a significant impact on thermal comfort. In contrast, the UTCI difference between sunlit and shaded areas during the spring equinox was smaller, likely due to weaker solar radiation and moderate temperatures in spring (Figure 6a).

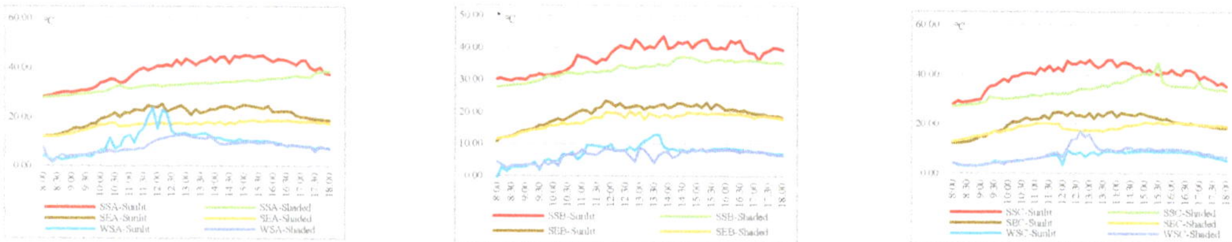

(**a**). Measured UTCI values at site A (**b**). Measured UTCI values at site B (**c**). Measured UTCI values at site C

Figure 6. Measured UTCI changes.

For site B during the winter solstice, the UTCI for the sunlit area was significantly higher than for the shaded area, but the overall values were lower than those for site A, indicating more limited sunlight conditions in winter. During the summer solstice, the UTCI for the sunlit area was higher than for the shaded area, though the difference was smaller than for site A, suggesting that site layout has a relatively smaller impact on thermal comfort. During the spring equinox, the UTCI for the sunlit area was still slightly higher than for the shaded area, but the difference was not significant (Figure 6b).

For site C during the winter solstice, the UTCI for the sunlit area was slightly lower than for the shaded area, which may be due to the shading and insulation effects of nearby trees. During the summer solstice, the UTCI for the sunlit area was the highest, much higher than for the shaded area, and was overall the highest among the three sites. The UTCI difference between the sunlit and shaded areas was smallest during the spring equinox, reflecting a more balanced distribution of sunlight and shadow (Figure 6c). However, the UTCI for the sunlit area during the winter solstice was slightly lower than for the shaded area, which could be related to the tree canopy in site C and the dense vegetation surrounding the site. Previous studies have shown that trees not only have a cooling effect [24], but also contribute to insulation during winter through various mechanisms. Thus, multiple uncertain factors may interfere with the site conditions during field surveys.

(3) Results and Analysis

First, the field measurement results confirmed the regulatory role of shadow on thermal comfort. The differences in UTCI values for shaded areas across different sites highlight the critical importance of site layout and shading design in optimizing thermal comfort.

Additionally, a clear correlation between UTCI differences and seasonal changes was observed. The gap in UTCI values between sunlit and shaded areas varied significantly with the seasons, being most pronounced during the summer and winter solstices and relatively balanced during the spring equinox.

During the summer solstice, the UTCI in sunlit areas was substantially higher than in shaded areas across all sites, demonstrating the direct impact of solar intensity on thermal comfort. Notably, site C recorded the highest UTCI values, likely due to inadequate shadow coverage provided by surrounding buildings. Similarly, during the winter solstice, the UTCI in sunlit areas was significantly higher than in shaded areas, particularly at site A and site B, highlighting the critical role of sunlight in improving thermal comfort during colder seasons.

3.2. Correlation Analysis Between Measured ISR and UTCI

The sunlight and shadow conditions recorded in the surveyed images were converted into hourly ISR data. This conversion facilitated subsequent discussions regarding the relationship between outdoor thermal comfort and ISR (Figure 7). Additionally, over 600 valid datasets were collected from both sunlit and shaded areas at each of the three measurement sites. Spearman's method was used for the correlation analysis.

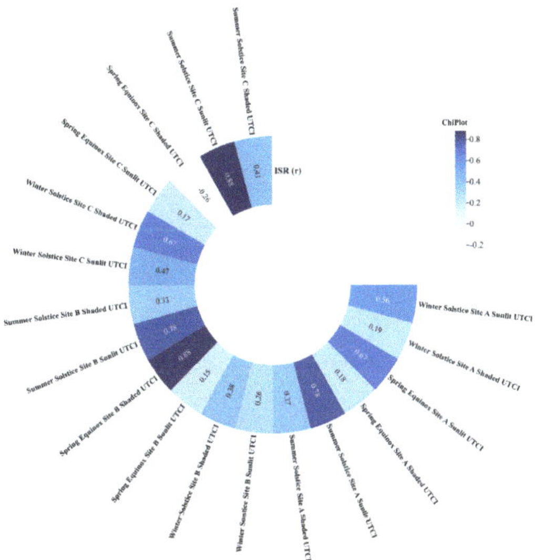

Figure 7. Correlation between ISR and seasonal comfort across three sites (source: calculated using SPSS).

A correlation analysis between the ISR and the UTCI was conducted for site A based on the measured data. The results showed that the UTCI in the sunlit areas exhibited a generally high degree of positive correlation with the ISR during the winter solstice, the spring equinox, and the summer solstice (Spearman's $|r| > 0.5$, $p < 0.01$). In contrast, the UTCI in the shaded areas showed a generally lower degree of positive correlation with the ISR. The analysis results for both sunlit and shaded areas indicated statistical significance.

For site B, the correlation analysis revealed a low degree of positive correlation between the UTCI and the ISR in the sunlit areas during the winter solstice and the spring equinox, while a higher degree of positive correlation was observed in the sunlit area during the summer solstice. In the shaded areas, the UTCI showed a high degree of positive correlation with ISR during the spring equinox.

For site C, the analysis indicated a moderate degree of positive correlation between the UTCI and the ISR in the sunlit areas during the winter solstice, a high degree of positive correlation during the summer solstice, and a low degree of positive correlation during the spring equinox. In the shaded areas, the UTCI exhibited a high degree of positive correlation with the ISR during the winter solstice, a moderate degree of positive correlation during the summer solstice, and a low degree of negative correlation during the spring equinox. Notably, site C was the only site where a negative correlation was observed.

3.3. Simulation Results

In traditional Chinese meteorology, the start of each season is marked by specific solar terms: Lichun (Beginning of Spring, 4 or 5 February), Lixia (Beginning of Summer, 5 or 6 May), Liqiu (Beginning of Autumn, 7 or 8 August), and Lidong (Beginning of Winter, 7 or 8 November). Following this seasonal division, the simulation periods in this study were defined as follows: spring (5 February to 5 May), summer (6 May to 7 August), autumn (8 August to 7 November), and winter (8 November to 4 February). This study primarily investigates the influence of solar shadow variations on outdoor thermal comfort. Therefore, the research period was set between 8:00 and 18:00, corresponding to daytime hours with direct solar exposure. The detailed simulation results are summarized in Table 2.

Table 2. Average simulated ISR values for the three sites across seasonal reference days.

Reference Day	Site A (Average ISR)	Site B (Average ISR)	Site C (Average ISR)
Winter Solstice (22 December)	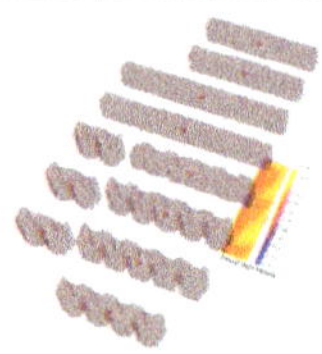 0.5887	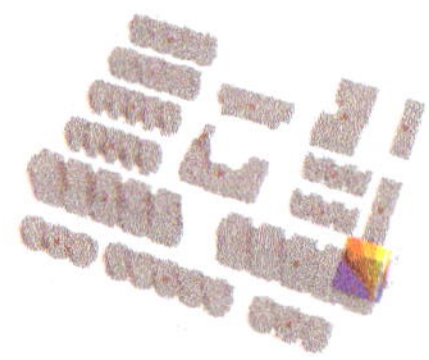 0.2399	 0.3107
Spring Equinox (21 March)	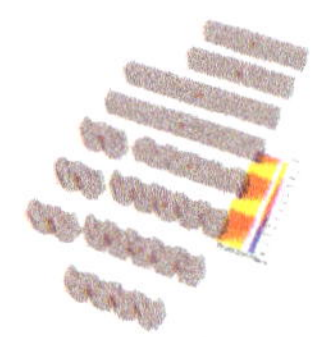 0.7152	 0.4615	 0.7427
Summer Solstice (22 June)	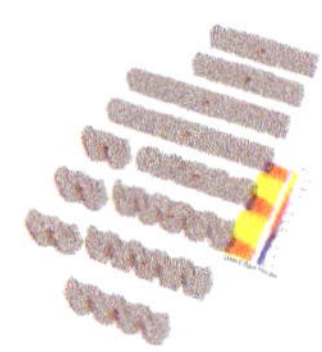 0.7815	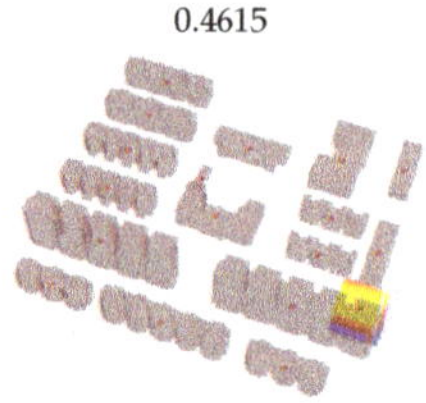 0.7731	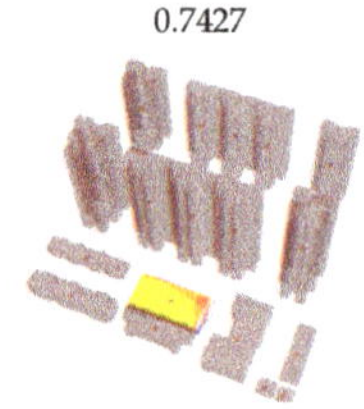 0.945
Autumn Equinox (23 September)	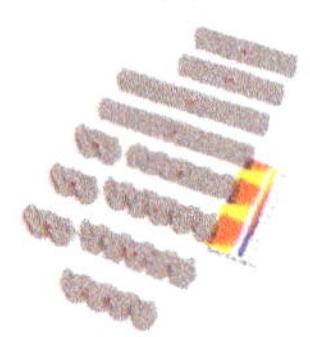 0.6999	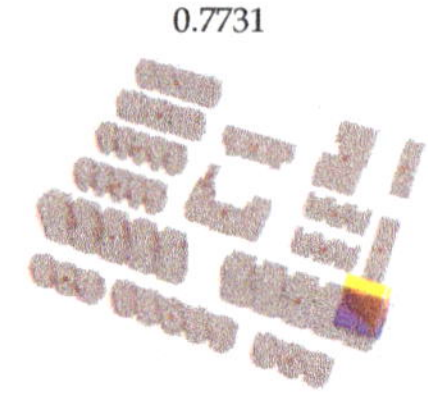 0.4516	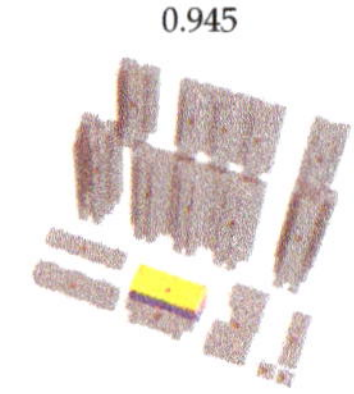 0.7507

3.3.1. Simulated ISR

Simulation of the ISR was conducted using 3D models of sites A, B, and C, representing different architectural typologies and spatial enclosures. The models were analyzed in Grasshopper with the Ladybug plugin to compute the hourly ISR values for each site during the defined seasonal periods.

Local geographic and climatic data, including latitude, longitude, solar angles, and weather files specific to Xi'an, were incorporated into the simulation. For each season, the insolation–shadow variation was calculated under clear sky conditions from 8:00 to 18:00. The simulated ISR values provided insights into the spatial and temporal distribution of sunlight and shadow across the study sites, which were subsequently analyzed for their effects on outdoor thermal comfort.

Through simulated calculations of the ISR for sites A, B, and C during the four typical solar terms, distinct seasonal trends and site-specific differences were observed. The specific changes in ISR trends are illustrated in Figure 8. At site A, ISR remained consistently high from 8:00 to 12:00 during all four seasons, showing minimal variation.

However, a significant decrease was observed after 12:00, with a pronounced decline in magnitude. At site B, the ISR exhibited a rapid decrease from 8:00 to 13:00 during the winter solstice, spring equinox, and autumnal equinox, followed by a relatively stable trend after 13:00, with minimal differences between time intervals. Conversely, during the summer solstice, the ISR at site B demonstrated a unique pattern, characterized by an initial decrease, a subsequent increase, and another decline, differing markedly from other seasonal trends. At site C, the ISR displayed a parabolic pattern, reaching its lowest values at 8:00 in the morning and 18:00 in the evening, exhibiting a symmetrical distribution.

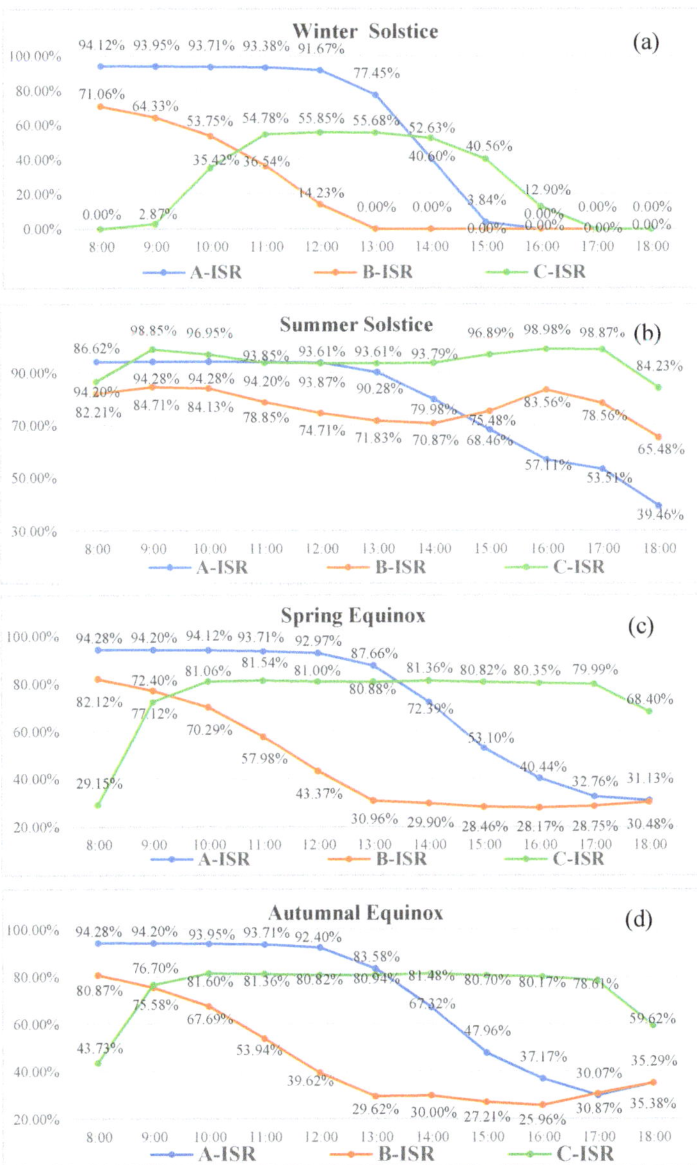

Figure 8. Simulated ISR dynamics for different sites. (**a**) Winter solstice; (**b**) Summer solstice; (**c**) Autumnal equinox; (**d**) Spring equinox.

3.3.2. Simulated UTCI

In the simulation process for the three sites, the shading effects of surrounding buildings and the variations in climatic conditions at different time points were considered. Illustrations of the simulation models for the three sites are presented in Table 3.

Table 3. Average simulated UTCI for the three sites during the simulated periods.

Reference Day	Site A (UTCI-Mean)	Site B (UTCI-Mean)	Site C (UTCI-Mean)
Winter Solstice (22 December)	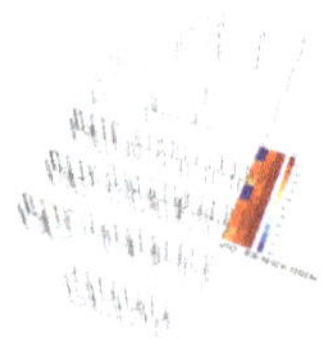 1.32–2.27 °C	 1.32–2.18 °C	 1.62–1.92 °C
Spring Equinox (21 March)	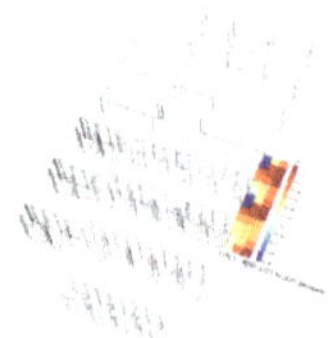 9.14–14.31 °C	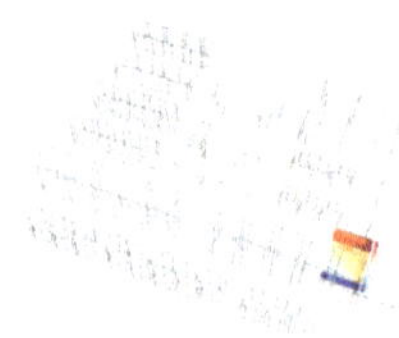 9.14–13.89 °C	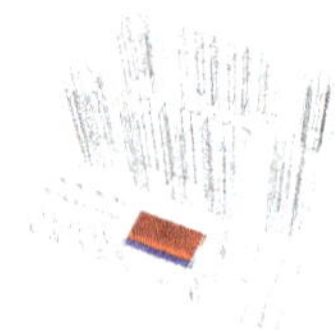 10.24–13.17 °C
Summer Solstice (22 June)	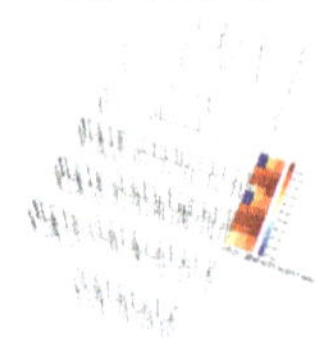 27.02–32.82 °C	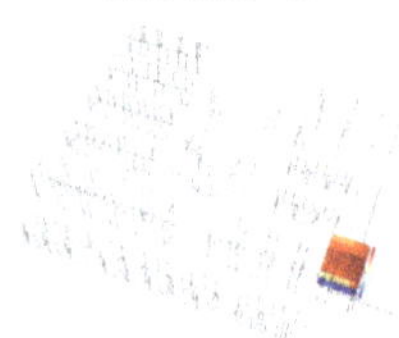 27.02–32.53 °C	 31.11–31.90 °C
Autumn Equinox (23 September)	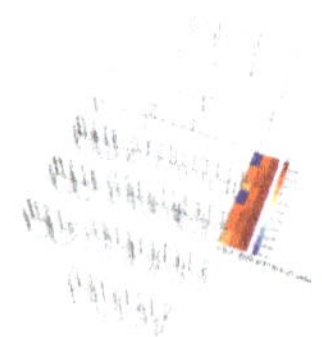 22.47–24.52 °C	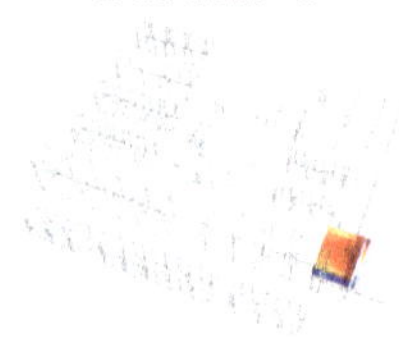 22.47–24.29 °C	 23.14–23.78 °C

(1) Comparison of simulated UTCI data

For the winter solstice, the simulated UTCI data for the three sites showed a trend of increasing followed by a decrease, peaking at 14:00 before declining. Site A exhibited the largest fluctuations between the minimum and maximum UTCI values. At 14:00, the minimum UTCI for site A rose to 3.1 °C, while the maximum reached 7.14 °C, demonstrating a significant increase in temperature difference. Site B's UTCI displayed a minimum of 3.1 °C and a maximum of 6.66 °C, slightly lower than site A's corresponding values. Site C's UTCI ranged from a minimum of 4.39 °C to a maximum of 5.65 °C, indicating less fluctuation in the thermal environment during the winter solstice.

For the summer solstice, the UTCI results for the three sites exhibited a steady increase, peaking at 14:00, after which they began to decline. At 14:00, the minimum UTCI for site A rose to 28.32 °C, with a maximum value of 34.19 °C, showing the largest difference between minimum and maximum values among the three sites. Site B followed a similar trend to site A, but during the peak high-temperature period (around 14:00), the maximum value was lower, reaching 33.88 °C. Site C displayed a similar trend in UTCI to sites A and B, but with overall lower temperature levels. At 14:00, site C's maximum UTCI was 33.2 °C, significantly lower than site A (34.19 °C) and site B (33.88 °C), indicating that its thermal environment performed better overall.

For both the spring equinox and the autumn equinox, the UTCI trend at site A was similar, with temperatures reaching their highest values around midday and a small day–night temperature difference, reflecting relatively stable thermal comfort during the transitional periods of the seasons. At site B, the UTCI also reached a high value around midday, with a small day–night temperature difference, but there was some divergence from site A, likely due to differences in shading layout and microclimate effects. At site C, the UTCI exhibited a characteristic midday high and a small day–night temperature difference. The values at different time points showed certain differences compared to sites A and B, reflecting the unique local conditions that affected the site.

(2) Results and Analysis

Based on the simulated UTCI data and visual analysis of the three sites during the four representative solar terms (Figure 9), it was observed that site A exhibited the largest threshold range between UTCI-Max and UTCI-Min. This indicates that site A experienced the most significant thermal environment fluctuations, primarily influenced by the shading effect of buildings on the western side of the site, which created considerable differences in solar and shaded areas and subsequently impacted thermal comfort.

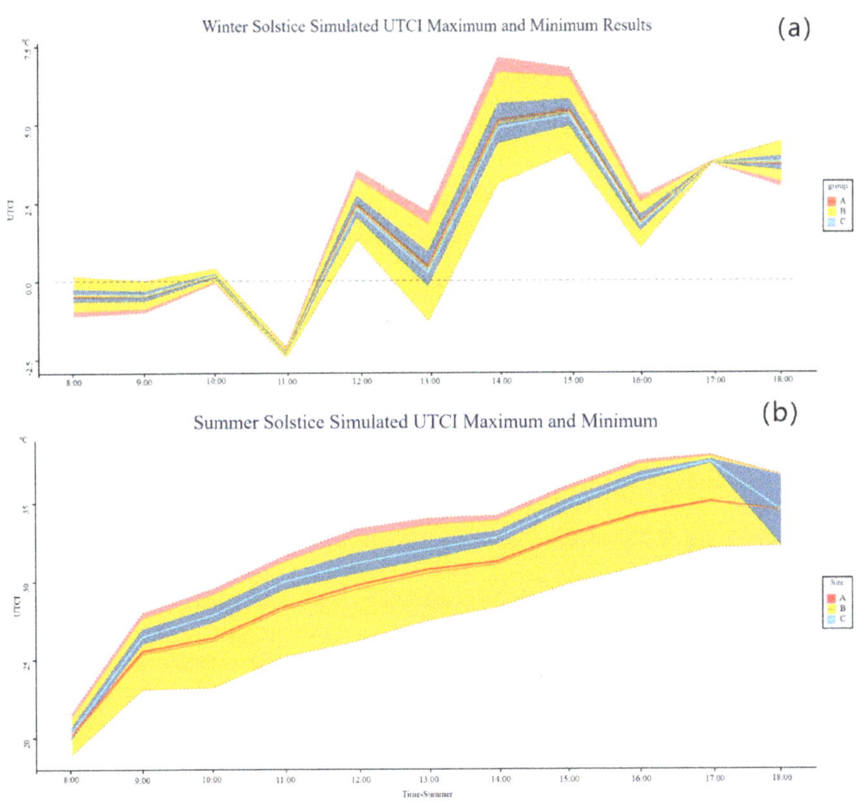

Figure 9. *Cont.*

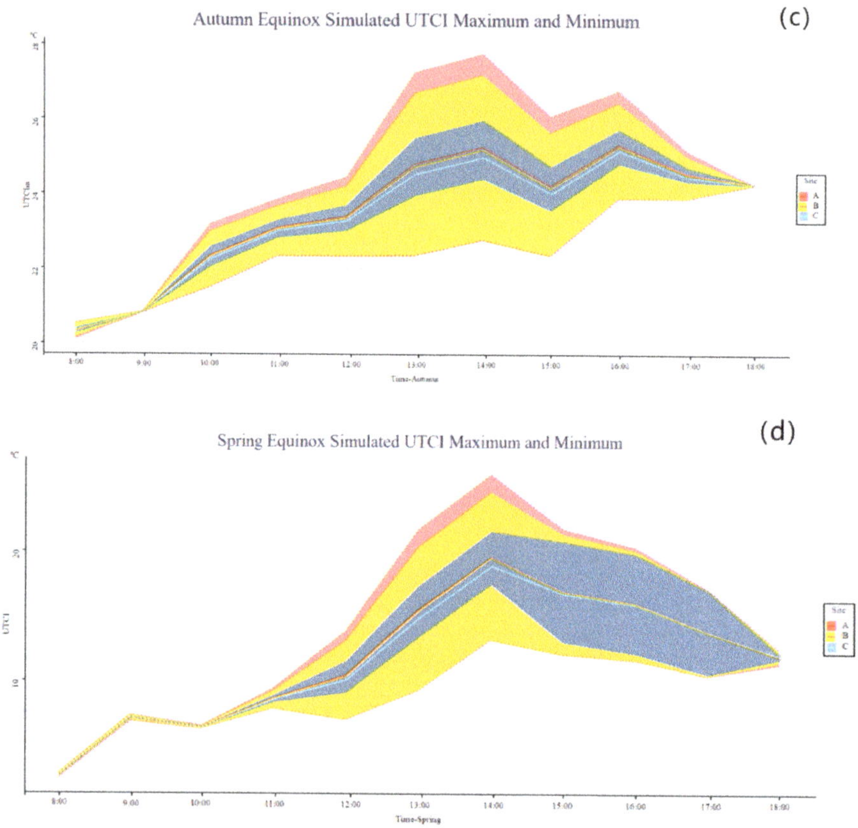

Figure 9. Simulated maximum and minimum UTCI values for different sites. (**a**) Winter solstice; (**b**) Summer solstice; (**c**) Autumnal equinox; (**d**) Spring equinox.

Site B showed a moderate range of UTCI differences, with shading effects from buildings on the northern and southern sides exerting a balanced impact on the thermal environment. Consequently, temperature variations within site B are less intense compared to site A but still exhibited a degree of unevenness.

Site C demonstrated the smallest range of UTCI differences. The layout of buildings on its southern and northern sides ensured a relatively stable distribution of ISR across the site, resulting in more consistent thermal environment conditions.

3.4. Workflow Efficiency Comparison

Comparative analysis shows that, for a typical 100 m × 100 m block, the ISR workflow (Grasshopper + Ladybug) completes geometry import and sun–shade setup in 2 min and computes seasonal hourly ISR in under 10 min on a standard desktop (Intel i7, 16 GB RAM). By contrast, an equivalent ENVI-met simulation requires approximately 1 h for setup and 3–4 h for runtime. Implementing ISR demands only intermediate parametric-design skills, whereas full microclimate modeling requires advanced CFD expertise. Overall, ISR reduces total project time from 6 h to under 30 min (over 90% time savings), decreases computational load by an order of magnitude, and lowers the technical barrier to thermal–environment optimization for practitioners.

3.5. Correlation Analysis Between Simulated ISR and UTCI

Using a simulation platform and locally available meteorological data, simulations were conducted with an hourly time step for ISR and UTCI during all four seasons. The simulation periods spanned from 8:00 to 18:00 daily, generating 900 to over 1000 datasets per site per season. After conducting normality tests on the data, it was observed that the data did not follow a normal distribution. Consequently, the Spearman algorithm was applied to analyze the correlation between the ISR and UTCI simulation results (Figure 10).

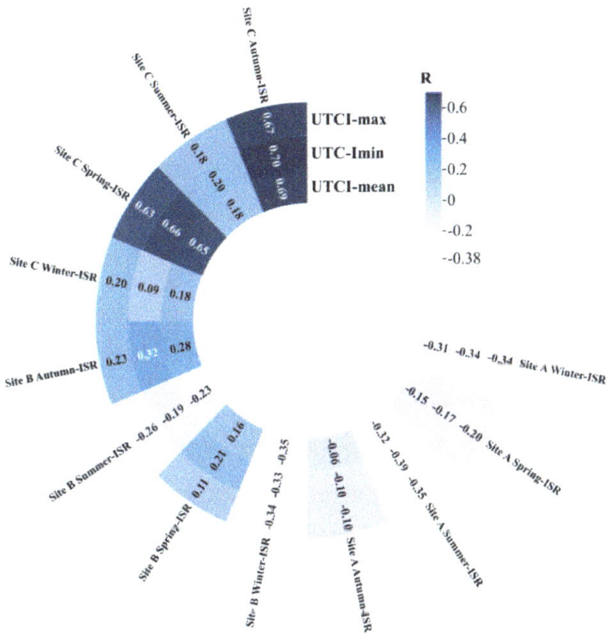

Figure 10. Correlation between simulated ISR and UTCI values across the three sites.

The relationship between the ISR and UTCI was examined during the simulation stage. An analysis of the simulation data for the three sites revealed that ISR plays a significant role in influencing UTCI, exerting both positive and negative effects under different conditions.

First, for site A, ISR showed varying degrees of negative correlation with the maximum, minimum, and mean UTCI values in all four seasons, with $p < 0.01$ indicating significance between ISR and UTCI. Second, for site B, ISR was correlated with the UTCI metrics for all four seasons. Finally, for site C, ISR was strongly positively correlated with UTCI metrics for the spring and fall seasons.

3.6. Sensitivity Analysis Between Simulated ISR and UTCI Values

The Sobol variance decomposition framework revealed ISR as the principal deterministic factor governing UTCI variations across all observational stations and seasonal cycles. Quantitative sensitivity metrics demonstrated exceptional dominance (Table 4). The sensitivity indices (Si and ST) of the ISR to the UTCI were close to or slightly above 1 for all sites and seasons, indicating that ISR is the most critical influence on UTCI, and its independent contribution covers almost the entire output variance. The values of Si and ST for all seasons and sites are almost equal (difference < 0.001), indicating that the effect of the ISR on the UTCI is dominated by independent effects, and the interaction effect is negligible; thus, it can be stated that the ISR has a dominant effect on the UTCI.

Table 4. Sensitivity analysis of different seasons between simulated ISR and UTCI values.

	Site A UTCI		Site B UTCI		Site C UTCI	
	Si	ST	Si	ST	Si	ST
Winter ISR	1.0388	1.0388	0.9943	0.9948	0.9715	0.9711
Summer ISR	0.9974	0.9974	0.9871	0.9871	1.0030	1.0030
Spring ISR	1.0088	1.0088	0.9834	0.9835	1.0030	1.0030
Autumn ISR	1.0060	1.0060	0.9994	0.9986	0.9808	0.9808

4. Discussion

4.1. *Effect of ISR Variation on the UTCI*

In different seasons, the interaction between the ISR and solar radiation intensity produces varying impacts on thermal comfort. Specifically, an increase in the ISR leads to greater solar radiation exposure in outdoor spaces, resulting in higher temperatures. While this effect improves thermal comfort in winter by warming the environment, it can cause overheating during summer, negatively affecting outdoor comfort.

Comparing the simulated results of the ISR and the UTCI across the three sites during four seasons, site A exhibited the largest ISR fluctuations from morning to evening. Correspondingly, site A also demonstrated the greatest difference between UTCI maximum and minimum values among the three sites. Site B followed, with moderate ISR variation and UTCI fluctuations. By contrast, site C showed a parabolic trend in ISR variation throughout the day, with smaller differences between morning and evening ISR values. Consequently, the difference in UTCI maximum and minimum values for site C was the smallest among the three sites.

Larger ISR variation indicates significant changes in solar exposure conditions, which results in pronounced surface temperature fluctuations and, subsequently, substantial UTCI changes. Conversely, smaller ISR variations signify more stable shadowing conditions, leading to relatively steady surface temperatures and smaller UTCI fluctuations. The strong influence of the ISR on the UTCI underscores the role of solar radiation in altering surface radiation intensity, which in turn affects human-perceived thermal loads.

The three study sites presented distinct ISR–UTCI patterns due to differences in morphological layout and environmental context. Site A, characterized by a relatively open layout, experienced the highest diurnal variation in ISR and the widest UTCI range, indicating strong sensitivity to temporal solar exposure. In contrast, site C, with higher vegetation density and enclosed spatial form, exhibited limited ISR fluctuation and the most stable UTCI values. Site B represented an intermediate case, where the correlation was seasonally variable and spatially balanced.

These variations suggest that the effectiveness of shading and solar exposure on outdoor thermal comfort is not uniform but influenced by microclimatic modifiers such as vegetation, enclosure ratio, and surface material. The results underline the need for localized design strategies that account for morphological context when applying ISR-based thresholds.

4.2. *Reasons for Differences Between Measured Data and Simulated Data*

Comparison between measured and simulated data revealed discrepancies in correlation trends across the three sites. For site A, the measured correlation was the opposite of the simulated results. For site B, the measured data for the spring and autumn equinoxes aligned with the simulation, while the measured results for the winter and summer solstices exhibited opposite correlations. For site C, the measured and simulated correlations showed consistent trends overall. The higher correlation observed in the field measurements (r = 0.883) versus the parametric simulations (r = 0.686) can be attributed to several factors. First, the datasets differ in temporal coverage and resolution: the measured ISR and UTCI values were recorded over a single day at one-minute intervals, capturing fine-scale fluctuations, whereas simulated values span an entire season at one-hour intervals, yielding smoother, aggregated profiles. Second, the empirical data incorporate real-world perturbations—shading from mature vegetation, heterogeneous ground-surface emissivity, and transient wind gusts—that enhance momentary ISR–UTCI coupling but are not fully represented in the idealized simulation environment. Third, simulation boundary conditions (e.g., uniform sky models, steady wind fields) and assumptions about surface albedo and canopy structure necessarily simplify complex urban microclimates, which can attenuate modeled correlation strength. Nonetheless, both Spearman coefficients exceed 0.6, indicating robust, statistically significant relationships between the ISR and the UTCI in both contexts.

These inconsistencies between the measured and simulated data may be attributed to several factors:

(1) Nonlinear relationships: The relationship between ISR and thermal comfort is not strictly linear. For instance, excessive solar exposure can increase discomfort in high-temperature seasons, while enhancing comfort in colder seasons. This suggests that the relationship between thermal comfort and ISR is complex, influenced by multiple factors such as season, wind speed, and humidity.
(2) Localized microclimate effects: Factors such as the built environment, vegetation coverage, and surface materials can alter local microclimatic conditions, affecting thermal comfort. Even with the same ISR, variations in material thermal properties (e.g., absorption and reflectivity) can lead to differing surface temperatures, causing deviations in measured results.
(3) Real-world complexity vs. idealized models: Real-world sites often involve numerous interacting factors not accounted for in idealized models. While the simulation primarily considers building-induced shadowing effects, real environments introduce additional uncertainties and overlapping variables, leading to deviations from predicted outcomes.

4.3. Data-Driven ISR Threshold Determination

A multi-dimensional analytical framework (field measurements, simulation validation, and sensitivity tests) systematically identified seasons with significant ISR–UTCI correlations ($|r| > 0.5$). We subsequently employed generalized additive models (GAMs) to characterize their nonlinear relationships, aiming to (1) quantify the complex influence of solar–shade dynamics (ISR) on outdoor thermal comfort (UTCI) and (2) determine critical ISR thresholds corresponding to the UTCI comfort range (18–26 °C), providing actionable parameters for daylight regulation in outdoor design.

GAMs were selected for their capacity to model nonlinear responses through thin-plate regression splines (Figure 11), outperforming conventional linear regression in detecting threshold inflection points. This flexibility proves critical when analyzing the dynamic equilibrium between ISR-mediated shade cooling and solar accessibility effects on the UTCI. To guard against overfitting, all smooth terms in the GAM were fitted using penalized thin-plate regression splines with a restricted basis dimension (k = 4) and smoothing parameters selected via REML (as implemented in the mgcv package). This penalization framework automatically balances model flexibility against parsimony.

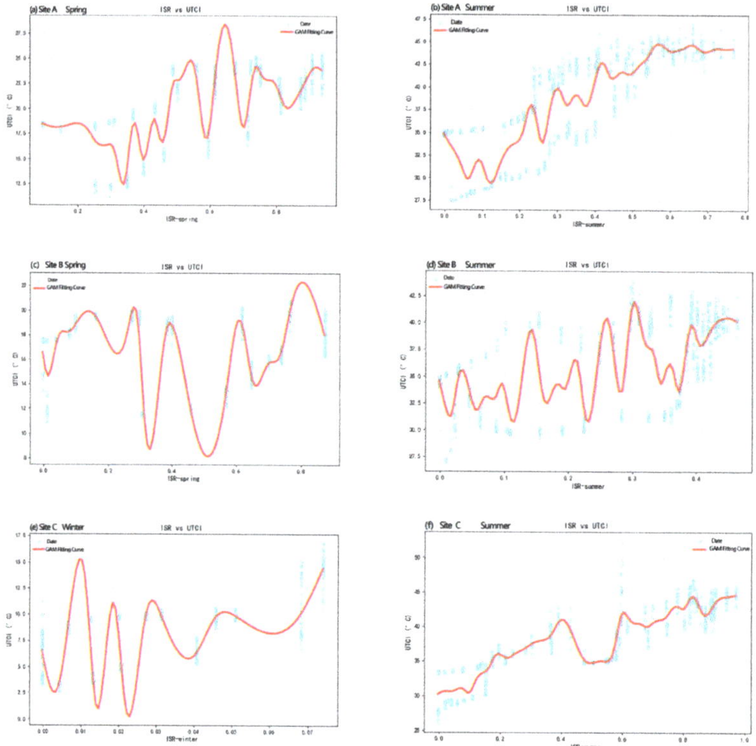

Figure 11. GAM model fitting based on measured data.

(1) Measured ISR Threshold Patterns

The results revealed pronounced seasonal variations in ISR comfort thresholds across sites: site A required 0.3423–0.6303 ISR during the vernal equinox, but ISR < 0.1243 at the summer solstice, while site B showed higher vernal thresholds (ISR: 0.5383–0.8030) with similar summer constraints (ISR < 0.1172). Site C necessitated winter ISR > 0.0554 to mitigate cold stress, contrasting sharply with near-zero ISR requirements (approaching 0) for summer thermal comfort attainment.

(2) Simulated ISR Threshold Patterns

Simulated data predictions further demonstrated(Figure 12) broader transitional season ISR thresholds (spring: 0.6465–0.8687; autumn: 0.0202–0.8384), a pattern fundamentally stemming from Xi'an's temperate monsoon climate.

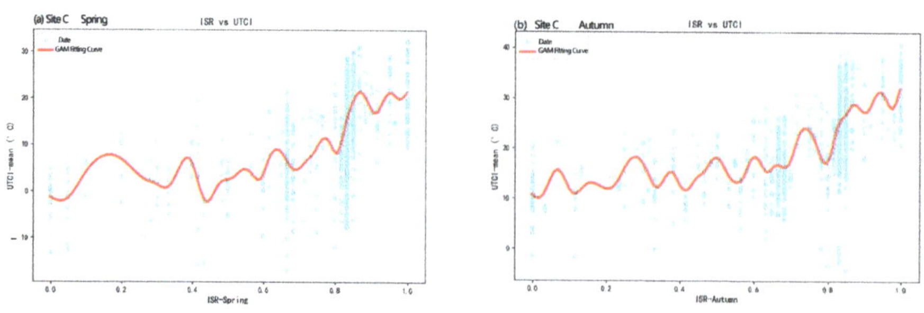

Figure 12. GAM model fitting based on simulated data.

The region's extreme thermal conditions—winter temperatures below −10 °C and summer peaks exceeding 40 °C—create significant deviations from the UTCI comfort range (9–26 °C). In contrast, spring or autumn mean temperatures (10–22 °C) align closely with UTCI criteria, enabling stable characterization of ISR–thermal comfort nonlinear relationships during these periods. Crucially, the findings demonstrate that extreme climates demand opposing design strategies: maximal shade coverage (ISR $\rightarrow$ 0) to counter summer heat stress versus minimal shading (ISR > 0.05) for winter solar gain optimization. While these results provide quantitative guidelines for seasonal dynamic shading design in transitional seasons, they simultaneously highlight limitations in winter–summer optimization frameworks, suggesting the need for auxiliary climate adaptation measures beyond ISR adjustments in extreme thermal scenarios.

4.4. Practical Implementation of ISR Thresholds

The Insolation Shadow Ratio (ISR) framework, while novel, is readily implementable through existing modeling platforms and simple guidelines. In the Xi'an residential case study, ISR thresholds (e.g., maintaining an hourly sunlit-area ratio $\geq$ 0.70 during 10 a.m.–2 p.m. at summer solstice) were derived via field measurements and Grasshopper + Ladybug simulations. Practitioners can replicate this workflow as follows.

Using Grasshopper, the LB_Sunlight Hourly and LB_ShadeAnalysis components can be combined to compute the ISR in real time, allowing building massing to be adjusted interactively until the model's hourly ISR meets the target threshold.

5. Conclusions

5.1. Limitations

This study exhibits three principal limitations requiring scholarly attention:

(a) Data Acquisition Constraints in Complex Urban Environments

Dense tree coverage in residential areas obstructed aerial views, preventing top-down capture of solar–shadow variations. Consequently, panoramic images from fixed ground vantage points were used, which may have constrained the spatial completeness and accuracy of ISR measurements. Future studies could employ drone-based photogrammetry or terrestrial LiDAR scanning to obtain high-resolution, three-dimensional shadow data. Additionally, resident activity diversity introduced variability into field observations, and differences in building orientation and height among the three hotspots limited direct comparability.

(b) Geographic Impact ISR Thresholds

The ISR metric exhibits strong spatial dependency, requiring new calculations in distinct geographical contexts due to its limited universal applicability.

(c) Mono-Variable Model Simplification

While the ISR-focused framework effectively reduces design complexity, it inherently neglects critical microclimate interactions. Future studies should integrate morphological and microclimatic parameters (e.g., humidity ratios, wind velocity) to better capture the multifactorial influences on UTCI. As this study represents the first systematic exploration of ISR as a morphological indicator, the thresholds reported herein reflect only the preliminary variability derived from UTCI. To enhance precision and establish more convergent, design-ready threshold ranges, future work will extend the GAM framework by incorporating covariates such as wind speed, relative humidity, air temperature, and solar radiation intensity.

(d) Lag Effects between Shading and Thermal Comfort Perception

While the ISR and the UTCI were synchronized at an hourly temporal resolution in this initial investigation, future research will include time-lagged cross-correlation analyses and higher-frequency simulations to capture any delayed thermal responses and improve the temporal robustness of the ISR–UTCI relationship.

Field measurements were conducted in late 2022 and early 2023, corresponding to the post-pandemic period in China. Although no lockdown measures were in place, and public activity had largely normalized, residual behavioral patterns may still have influenced outdoor space usage and thermal comfort responses. Future studies are encouraged to collect data under fully stabilized social conditions to further validate the representativeness of ISR–UTCI relationships. Although the ISR proved reproducible and highly correlated with the UTCI, a detailed sensitivity analysis was not undertaken. Our workflow relies on precise building and vegetation geometry to ensure repeatability but has yet to assess ISR's response to variations in height, aspect ratio, or canopy density. In addition, panoramic imagery—used solely for validation—introduces uncertainty through thresholding and spatial parameterization. Future work will (1) perform systematic parametric sweeps of urban form, (2) run perturbation tests to quantify imaging errors, and (3) develop robustness metrics (e.g., confidence intervals) to refine ISR thresholds across diverse morphologies.

5.2. Key Findings

This study introduces a novel morphological indicator, the Insolation Shadow Ratio (ISR), of outdoor spaces in residential areas. The relationship between the ISR and outdoor thermal comfort (UTCI) was explored, revealing the following findings:

(1) Analytical results of the measured data:

① A correlation between the ISR and the UTCI was observed across different sites, and it was validated using SPSS based on measured ISR and UTCI data from the actual residential area. The degree of correlation varied among the three sites. Site A and site B exhibited stronger positive correlations, while site C demonstrated a weaker correlation. An exception was observed during the vernal equinox at site C, where the measured ISR and UTCI data showed a negative correlation. Site C exhibited outliers during the vernal equinox. However, this anomaly is likely influenced by complex environmental factors, making it difficult to attribute the phenomenon to a single cause.

② The strength of the correlation between the ISR and the UTCI of site A and site C was significantly higher in the sunlit areas than in the shaded areas; however, at site B, it was higher in the shaded areas than in the sunlit areas. The stronger correlation between ISR and UTCI in shaded areas compared to sunlit areas can be explained by two main factors: environmental factors and measurement point distribution. Shaded areas are more influenced by factors like building layout, vegetation, wind speed, and humidity. For example, vegetation in shaded areas can further cool the environment through transpiration, enhancing the ISR–UTCI correlation. Sunlit areas, less affected by these factors, show a weaker correlation. The placement of measurement points and surrounding shading from buildings or vegetation can amplify the impact of ISR on UTCI in shaded areas, especially when these areas are heavily obstructed or influenced by microclimatic conditions. Therefore, this demonstrates that shaded areas are more sensitive to environmental and microclimatic interactions than sunlit areas.

(2) Analytical results of the simulation date:

The three sites all demonstrated a correlation between the ISR and the UTCI, but with varying degrees and trends (positive or negative).

① The ISRs of site A on the four typical solar term days showed negative correlations with the UTCI, of different degrees.
② The ISRs of site B showed positive correlations with the UTCI on the spring and autumn equinoxes and negative correlations on the winter and summer solstices.
③ The ISRs of site C on the four typical solar term days showed positive correlations with the UTCI, of different degrees.
④ The magnitude of the ISR variations had a direct impact on the range of outdoor thermal comfort fluctuations. As the ISR differences increased, the UTCI fluctuations became more pronounced, suggesting ISR as a potentially key factor in shaping the thermal environment of outdoor spaces.

Through this study path and the seasonal fitting results of the GAM (generalized additive model), spring and autumn can better find ISR threshold ranges, possibly because the UTCI corresponding comfort range is 9–26 °C. Xi'an is located in a temperate monsoon climate in the alpine region of the Chinese Plateau, with distinct seasons. The average temperature in cold and dry winters is 1 to −1 °C, and the minimum is below −10 °C; the highest temperature in summer heat is more than 40 °C. Therefore, the comfort range is obviously different from the actual temperature in winter and summer in Xi'an.

Although the ISR is unprecedented, the strong positive correlations we observe between higher sunlit fractions and improved UTCI values echo similar patterns reported for static sky view metrics [22], where increased sky exposure corresponded to elevated PET and UTCI in cold-season conditions. Likewise, our finding of inverted relationships under dense canopy at site C parallels Morganti et al.'s [2] demonstration that vegetation density can reverse expected comfort trends by prolonging low-angle shading. By situating ISR's dynamic radiative insights alongside these established static and vegetative studies, we underscore both its continuity with—and advancement beyond—prior morphological analyses.

Overall, the analytical results of both the measured and simulated data revealed a correlation between the ISR and the UTCI in some seasons (primary outcome at site C in spring and autumn) and patterns of the influence of ISR variations on UTCI fluctuations. Therefore, the ISR indicator could potentially be used as a tool in outdoor space design for improving the outdoor thermal environment in specific areas. Future research can be conducted to further clarify and validate the correlation between the ISR and the outdoor thermal environment under different climatic conditions, so as to develop a more comprehensive understanding of the ISR–thermal comfort relationship.

5.3. Contributions

This study makes three seminal contributions to climate-responsive urban design:

(1) Theoretical Innovation: We propose the ISR as a novel morphometric index that quantifies solar shade dynamics and their nonlinear coupling with UTCI-based thermal comfort. This metric establishes an intuitive visualization framework linking shadow coverage (0–1 normalized) to physiological heat stress thresholds (9–26 °C UTCI).
(2) Design Praxis Transformation: The ISR enables direct conversion of thermal comfort requirements into implementable shade design parameters and reduces the technical threshold of thermal environment optimization.
(3) Build A New Research Pathway: We establish a dynamic solar exposure analysis framework that integrates field microclimate measurements and simulated data with GAM-based threshold modeling. Although the present study focuses on Xi'an, the ISR computation protocol—grounded in local solar trajectories and parametric solar–shadow modeling—is directly transferable to other cold-region contexts.

Author Contributions: Conceptualization, J.S. and Y.L.; Methodology, J.S., Y.L. and S.C.; Software, J.S. and B.L.; Validation, J.S. and B.L.; Formal analysis, J.S., D.H.C.C. and S.C.; Investigation, J.S. and B.L.; Resources, Y.L. and D.H.C.C.; Data curation, D.H.C.C.; Writing—original draft, J.S. and Y.L.; Writing—review & editing, Y.L. and D.H.C.C.; Visualization, J.S.; Supervision, Y.L.; Funding acquisition, Y.L. All authors have read and agreed to the published version of the manuscript.

Funding: This work was funded by the National Natural Science Foundation of China (Project No. 52078422). This work was also funded by the Royal Society Funding Number: IEC\NSFC\233170-International Exchanges 2023 Cost Share (NSFC)-Research on key technical pathways for low-carbon construction of urban residential areas based on climate change assessment and vernacular examples.

Data Availability Statement: No new data were created for this study, if there is a request for data please contact the authors to provide it.

Acknowledgments: Declaration of generative AI and AI-assisted technologies in the writing process: during the preparation of this work, the authors used AI only for language polishing. The authors reviewed and edited the content as needed and take full responsibility for the content of the paper.

Conflicts of Interest: The authors declare no conflict of interest.

Appendix A

Table A1. Records of variations in sunshine and shadow at site A during the winter solstice.

22 December 2022	Panoramic Picture
8:00–10:00	
12:00–14:00	
16:00–18:00	

Table A2. Records of variations in sunshine and shadow at site A during the spring equinox.

21 March 2023	Panoramic Picture
8:00–10:00	
12:00–14:00	
16:00–18:00	

Table A3. Records of variations in sunshine and shadow at site A during the summer solstice.

22 June 2023	Panoramic Picture
8:00–10:00	
12:00–14:00	
16:00–18:00	

Table A4. Time-wise ISR simulation of the three sites.

	Time	Winter Solstice	Spring Equinox	Summer Solstice	Autumnal Equinox
Site A	8:00	0.941	0.943	0.942	0.943
	9:00	0.940	0.942	0.943	0.942
	10:00	0.937	0.941	0.943	0.940
	11.00	0.934	0.937	0.942	0.937
	12:00	0.917	0.930	0.939	0.924
	13:00	0.775	0.877	0.903	0.836
	14:00	0.406	0.724	0.800	0.673
	15:00	0.038	0.531	0.685	0.480
	16:00	0.000	0.404	0.571	0.372
	17:00	0.000	0.328	0.535	0.301
	18:00	0.000	0.311	0.395	0.353
Site B	8:00	0.711	0.821	0.822	0.809
	9:00	0.643	0.771	0.847	0.756
	10:00	0.538	0.703	0.841	0.677
	11:00	0.365	0.580	0.788	0.539
	12:00	0.142	0.434	0.747	0.396
	13:00	0.000	0.310	0.718	0.296
	14:00	0.000	0.299	0.709	0.300
	15:00	0.000	0.285	0.755	0.272
	16:00	0.000	0.282	0.836	0.260
	17:00	0.000	0.288	0.786	0.309
	18:00	0.000	0.305	0.655	0.354
Site C	8:00	0.000	0.292	0.866	0.437
	9:00	0.029	0.724	0.988	0.767
	10:00	0.354	0.811	0.970	0.816
	11:00	0.548	0.815	0.938	0.814
	12:00	0.559	0.810	0.936	0.808
	13:00	0.557	0.809	0.936	0.809
	14:00	0.526	0.814	0.938	0.815
	15:00	0.406	0.808	0.969	0.807
	16:00	0.129	0.803	0.990	0.802

Table A5. UTCI simulation at site A.

Time	Winter Solstice		Spring Equinox		Summer Solstice		Autumnal Equinox	
	Min	Max	Min	Max	Min	Max	Min	Max
8:00	−1.1	0.16	2.43	2.81	18.92	21.57	20.1	20.55
9:00	−0.99	0.02	6.87	7.38	23.08	28	20.83	20.87
10:00	−0.07	0.41	6.29	6.56	23.19	29.54	21.51	23.2
11:00	−2.45	−2.09	7.87	9.43	25.16	31.62	22.34	23.86
12:00	1.3	3.55	7.02	13.88	26.12	33.32	22.34	24.46
13:00	−1.28	2.23	9.27	21.77	27.45	34.01	22.36	27.26
14:00	3.1	7.14	13.21	26.01	28.32	34.19	22.78	27.75
15:00	4.07	6.8	12.09	21.79	29.81	36.14	22.36	26.09
16:00	1.04	2.71	11.65	20.38	30.87	37.73	23.89	26.78
17:00	3.73	3.81	10.46	17.19	32.09	38.09	23.95	25.18
18:00	2.98	4.44	11.38	12.49	32.21	36.79	24.26	24.27

Table A6. UTCI simulation at site B.

Time	Winter Solstice		Spring Equinox		Summer Solstice		Autumnal Equinox	
	Min	Max	Min	Max	Min	Max	Min	Max
8:00	−0.95	0.16	2.48	2.81	18.92	21.3	20.15	20.55
9:00	−0.88	0.02	6.93	7.38	23.08	27.66	20.83	20.87
10:00	−0.02	0.41	6.29	6.53	23.19	29.18	21.51	23.01
11:00	−2.45	−2.13	7.87	9.25	25.16	31.26	22.34	23.68
12:00	1.3	3.29	7.02	13.13	26.12	32.84	22.34	24.21
13:00	−1.28	1.82	9.27	20.41	27.45	33.57	22.36	26.71
14:00	3.1	6.66	13.21	24.66	28.32	33.88	22.78	27.19
15:00	4.07	6.49	12.09	21.42	29.81	35.87	22.36	25.66
16:00	1.04	2.52	11.65	20.17	30.87	37.49	23.89	26.45
17:00	3.73	3.8	10.46	17.14	32.09	38	23.89	25.04
18:00	3.15	4.44	11.51	12.49	32.21	36.78	24.26	24.27

Table A7. UTCI simulation at site C.

Time	Winter Solstice		Spring Equinox		Summer Solstice		Autumnal Equinox	
	Min	Max	Min	Max	Min	Max	Min	Max
8:00	−0.63	−0.24	2.57	2.69	19.94	20.7	20.27	20.4
9:00	−0.62	−0.31	7.06	7.21	25.99	26.92	20.84	20.85
10:00	0.1	0.25	6.38	6.46	27.36	28.37	22.06	22.59
11:00	−2.34	−2.22	8.37	8.86	29.48	30.47	22.83	23.3
12:00	2.02	2.72	9.12	11.51	30.49	31.8	23.02	23.69
13:00	−0.16	0.93	13.41	17.36	31.4	32.61	23.96	25.5
14:00	4.39	5.65	17.54	21.59	32.36	33.2	24.41	25.97
15:00	4.94	5.8	13.05	20.79	34.55	35.28	23.57	24.74
16:00	1.58	2.1	12.17	19.83	36.33	36.97	24.82	25.73
17:00	3.75	3.78	10.58	17.06	37.55	37.8	24.35	24.73
18:00	3.52	3.97	11.78	12.13	32.27	36.75	24.26	24.27

References

1. Huo, K.; Qin, R.; Zhao, J.; Ma, X. Long-Term Tracking of Urban Structure and Analysis of Its Impact on Urban Heat Stress: A Case Study of Xi'an, China. *Ecol. Indic.* **2025**, *174*, 113418. [CrossRef]
2. Morganti, M.; Salvati, A.; Coch, H.; Cecere, C. Urban Morphology Indicators for Solar Energy Analysis. *Energy Procedia* **2017**, *134*, 807–814. [CrossRef]

3. Open Knowledge Maps. Knowledge Map for Research on Outdoor Thermal Comfort in Shaded Areas. 11 June 2024. Available online: https://openknowledgemaps.org/map/b898def6d0ee7b3ff345d7fdfa13fb5c (accessed on 4 June 2024).
4. Lai, D.; Liu, W.; Gan, T.; Liu, K.; Chen, Q. A Review of Mitigating Strategies to Improve the Thermal Environment and Thermal Comfort in Urban Outdoor Spaces. *Sci. Total Environ.* **2019**, *661*, 337–353. [CrossRef]
5. Elnabawi, M.H.; Jamei, E. The thermal perception of outdoor urban spaces in a hot arid climate: A structural equation modelling (SEM) approach. *Urban Clim.* **2024**, *55*, 101969. [CrossRef]
6. Jänicke, B.; Milošević, D.; Manavvi, S. Review of User-Friendly Models to Improve the Urban Micro-Climate. *Atmosphere* **2021**, *12*, 1291. [CrossRef]
7. Lau, K.K.-L.; Lindberg, F.; Rayner, D.; Thorsson, S. The Effect of Urban Geometry on Mean Radiant Temperature under Future Climate Change: A Study of Three European Cities. *Int. J. Biometeorol.* **2015**, *59*, 799–814. [CrossRef]
8. Li, J. A Theory and Method of Solar Adaptive Urban Design. Ph.D. Thesis, Southeast University, Nanjing, China, 2019. Available online: https://kns.cnki.net/kcms2/article/abstract (accessed on 8 June 2024).
9. Compagnon, R. Solar and Daylight Availability in the Urban Fabric. *Energy Build.* **2004**, *36*, 321–328. [CrossRef]
10. Salata, F.; Golasi, I.; de Lieto Vollaro, R.; de Lieto Vollaro, A. Urban Microclimate and Outdoor Thermal Comfort. A Proper Procedure to Fit ENVI-Met Simulation Outputs to Experimental Data. *Sustain. Cities Soc.* **2016**, *26*, 318–343. [CrossRef]
11. Acero, J.A.; Koh, E.J.Y.; Tan, Y.S.; Leslie; Norford, K. Quantifying the Effect of Building Shadowing and Cloudiness on Mean Radiant Temperature in Singapore. *Atmosphere* **2021**, *12*, 1012. [CrossRef]
12. Wang, M.; Gou, Z. Gaussian Mixture Model Based Classification for Analyzing Longitudinal Outdoor Thermal Environment Data to Evaluate Comfort Conditions in Urban Open Spaces. *Urban Clim.* **2024**, *53*, 101792. [CrossRef]
13. Huang, T.; Li, J.; Xie, Y.; Niu, J.; Mak, C.M. Simultaneous Environmental Parameter Monitoring and Human Subject Survey Regarding Outdoor Thermal Comfort and Its Modelling. *Build. Environ.* **2017**, *125*, 502–514. [CrossRef]
14. Johansson, E.; Emmanuel, R. The Influence of Urban Design on Outdoor Thermal Comfort in the Hot, Humid City of Colombo, Sri Lanka. *Int. J. Biometeorol.* **2006**, *51*, 119–133. [CrossRef] [PubMed]
15. Song, Z.; Wang, Y.; Li, J.; Liu, T.; Li, Y. How Building Arrangements Can Improve Outdoor Thermal Comfort and Indoor Sunlight? *Proc. Inst. Civ. Eng.—Munic. Eng.* **2024**, *177*, 111–129. [CrossRef]
16. Liu, Z.; Li, J.; Xi, T. A Review of Thermal Comfort Evaluation and Improvement in Urban Outdoor Spaces. *Buildings* **2023**, *13*, 3050. [CrossRef]
17. Costa, I.T.; Wollmann, C.A.; Writzl, L.; Iensse, A.C.; da Silva, A.N.; de Freitas Baumhardt, O.; Gobo, J.P.A.; Shooshtarian, S.; Matzarakis, A. A Systematic Review on Human Thermal Comfort and Methodologies for Evaluating Urban Morphology in Outdoor Spaces. *Climate* **2024**, *12*, 30. [CrossRef]
18. Han, Y.; Jo, Y.; Kim, E.J. Influence of Landscape Interventions on Thermal Comfort under Time-Varying Building Shadow; New Gwanghwamun Square Case, Seoul, South Korea. *Heliyon* **2023**, *9*, e19436. [CrossRef]
19. Liu, B.; Liu, Y.; Cho, S.; Chow, D.H.C. Urban Morphology Indicators and Solar Radiation Acquisition: 2011–2022 Review. *Renew. Sustain. Energy Rev.* **2024**, *199*, 114548. [CrossRef]
20. Nouri, A.S.; Costa, J.P.; Matzarakis, A. Examining Default Urban-Aspect-Ratios and Sky-View-Factors to Identify Priorities for Thermal-Sensitive Public Space Design in Hot-Summer Mediterranean Climates: The Lisbon Case. *Build. Environ.* **2017**, *126*, 442–456. [CrossRef]
21. Miao, C.; Yu, S.; Hu, Y.; Zhang, H.; He, X.; Chen, W. Review of Methods Used to Estimate the Sky View Factor in Urban Street Canyons. *Build. Environ.* **2020**, *168*, 106497. [CrossRef]
22. Watson, I.D.; Johnson, G.T. Graphical Estimation of Sky View-Factors in Urban Environments. *J. Climatol.* **1987**, *7*, 193–197. [CrossRef]
23. Yi, P.; Liu, L.; Huang, Y.; Zhang, M.; Liu, H.; Bedra, K.B. Study on the Coupling Relationship between Thermal Comfort and Urban Center Spatial Morphology in Summer. *Sustainability* **2023**, *15*, 5084. [CrossRef]
24. Yan, H.; Zhang, L.; Ding, X.; Zhang, Z.; Qi, Z.; Jiang, L.; Bu, D. Multi-Objective Optimization of Outdoor Thermal Comfort and Sunlight Hours in Elderly Residential Areas: A Case Study of Beijing, China. *Buildings* **2024**, *14*, 3770. [CrossRef]
25. Zhang, M.; You, W.; Qin, Q.; Peng, D.; Hu, Y.; Gao, Z.; Buccolieri, R. Investigation of Typical Residential Block Typologies and Their Impact on Pedestrian-Level Microclimate in Summers in Nanjing, China. *Front. Archit. Res.* **2022**, *11*, 278–296. [CrossRef]
26. Salgado-Conrado, L.; Lopez-Montelongo, A.; Alvarez-Macias, C.; Hernadez-Jaquez, J. Review of Heliodon Developments and Computational Tools for Building Shadow Analysis. *Buildings* **2022**, *12*, 627. [CrossRef]
27. Lin, T.-P.; Matzarakis, A.; Hwang, R.-L. Shading Effect on Long-Term Outdoor Thermal Comfort. *Build. Environ.* **2010**, *45*, 213–221. [CrossRef]
28. Lindberg, F.; Grimmond, C.S.B. Nature of vegetation and building morphology characteristics across a city: Influence on shadow patterns and mean radiant temperatures in London. *Urban Ecosyst.* **2011**, *14*, 617–634. [CrossRef]
29. Memon, R.A.; Chirarattananon, S.; Vangtook, P. Thermal Comfort Assessment and Application of Radiant Cooling: A Case Study. *Build. Environ.* **2008**, *43*, 1185–1196. [CrossRef]

30. de Dear, R.J.; Brager, G.S. Thermal Comfort in Naturally Ventilated Buildings: Revisions to ASHRAE Standard 55. *Energy Build.* **2002**, *34*, 549–561. [CrossRef]
31. Shaw, E.W. Thermal Comfort: Analysis and Applications in Environmental Engineering, by P. O. Fanger. 244 Pp. DANISH TECHNICAL PRESS. Copenhagen, Denmark, 1970. Danish Kr. 76, 50. *R. Soc. Health J.* **1972**, *92*, 164. [CrossRef]
32. Das, S.; Subudhi, S. A Review on Different Methodologies to Study Thermal Comfort. *Int. J. Environ. Sci. Technol.* **2022**, *19*, 2155–2171. [CrossRef]
33. Lai, D.; Chen, B.; Liu, K. Quantification of the Influence of Thermal Comfort and Life Patterns on Outdoor Space Activities. *Build. Simul.* **2020**, *13*, 113–125. [CrossRef]
34. Höppe, P. The Physiological Equivalent Temperature—A Universal Index for the Biometeorological Assessment of the Thermal Environment. *Int. J. Biometeorol.* **1999**, *43*, 71–75. [CrossRef]
35. Matzarakis, A.; Mayer, H.; Iziomon, M.G. Applications of a Universal Thermal Index: Physiological Equivalent Temperature. *Int. J. Biometeorol.* **1999**, *43*, 76–84. [CrossRef] [PubMed]
36. Ji, W.; Zhu, Y.; Du, H.; Cao, B.; Lian, Z.; Geng, Y.; Liu, S.; Xiong, J.; Yang, C. Interpretation of Standard Effective Temperature (SET) and Explorations on Its Modification and Development. *Build. Environ.* **2022**, *210*, 108714. [CrossRef]
37. Jia, X.; Li, S.; Zhu, Y.; Du, Z.; Cao, B. Development of the Universal Standard Effective Temperature for Evaluating Thermal Comfort across Different Metabolic Rates. *Build. Environ.* **2024**, *250*, 111149. [CrossRef]
38. Zhao, Q.; Lian, Z.; Lai, D. Thermal Comfort Models and Their Developments: A Review. *Energy Built Environ.* **2021**, *2*, 21–33. [CrossRef]
39. Bröde, P.; Krüger, E.; Fiala, D. UTCI: Validation and Practical Application to the Assessment of Urban Outdoor Thermal Comfort. *Geogr. Pol.* **2013**, *86*, 11–20. [CrossRef]
40. Fiala, D.; Havenith, G.; Broede, P.; Kampmann, B.; Jendritzky, G. UTCI-Fiala Multi-Node Model of Human Heat Transfer and Temperature Regulation. *Int. J. Biometeorol.* **2012**, *56*, 429–441. [CrossRef]
41. Mahdavinejad, M.; Shaeri, J.; Nezami, A.; Goharian, A. Comparing Universal Thermal Climate Index (UTCI) with Selected Thermal Indices to Evaluate Outdoor Thermal Comfort in Traditional Courtyards with BWh Climate. *Urban Clim.* **2024**, *54*, 101839. [CrossRef]
42. Wereski, S.; Krzyżewska, A.; Dobek, M. Winter UTCI Variability in Poland in the 21st Century. *Misc. Geogr.* **2020**, *24*, 128–137. [CrossRef]
43. Blazejczyk, K.; Epstein, Y.; Jendritzky, G.; Staiger, H.; Tinz, B. Comparison of UTCI to Selected Thermal Indices. *Int. J. Biometeorol.* **2012**, *56*, 515–535. [CrossRef] [PubMed]
44. Zare, S.; Hasheminejad, N.; Shirvan, H.E.; Hemmatjo, R.; Sarebanzadeh, K.; Ahmadi, S. Comparing Universal Thermal Climate Index (UTCI) with Selected Thermal Indices/Environmental Parameters during 12 Months of the Year. *Weather. Clim. Extrem.* **2018**, *19*, 49–57. [CrossRef]
45. Jin, H.; Liu, S.; Kang, J. Thermal Comfort Range and Influence Factor of Urban Pedestrian Streets in Severe Cold Regions. *Energy Build.* **2019**, *198*, 197–206. [CrossRef]
46. Nie, T.; Lai, D.; Liu, K.; Lian, Z.; Yuan, Y.; Sun, L. Discussion on Inapplicability of Universal Thermal Climate Index (UTCI) for Outdoor Thermal Comfort in Cold Region. *Urban Clim.* **2022**, *46*, 101304. [CrossRef]
47. Wu, J.; Jin, T.; Wu, Y.; Ding, Y.; Mu, Y.; Zeng, D. The Variation of UTCI with the Background of Climate Change and Its Implications for Tourism in a Complicated Climate Region in Western China. *Sustainability* **2022**, *14*, 15047. [CrossRef]
48. Wu, Z.; Kong, F.; Wang, Y.; Sun, R.; Chen, L. The Impact of Greenspace on Thermal Comfort in a Residential Quarter of Beijing, China. *Int. J. Environ. Res. Public Health* **2016**, *13*, 1217. [CrossRef]
49. Park, S.; Stanton; Tuller, E.; Jo, M. Application of Universal Thermal Climate Index (UTCI) for Microclimatic Analysis in Urban Thermal Environments. *Landsc. Urban Plan.* **2014**, *125*, 146–155. [CrossRef]
50. Ning, S.; Jing, W.; Ge, Z. Sunlight Perception and Outdoor Thermal Comfort in College Campuses: A New Perspective. *Sci. Rep.* **2023**, *13*, 16112. [CrossRef]
51. Cohen, J.; Cohen, P.; Stephen; West, G.; Aiken, L. *Applied Multiple Regression/Correlation Analysis for the Behavioral Sciences*, 3rd ed.; Routledge: New York, NY, USA, 2013. [CrossRef]
52. Jebli, I.; Belouadha, F.-Z.; Kabbaj, M.I.; Tilioua, A. Prediction of Solar Energy Guided by Pearson Correlation Using Machine Learning. *Energy* **2021**, *224*, 120109. [CrossRef]
53. de Winter, J.C.F.; Gosling, S.D.; Potter, J. Comparing the Pearson and Spearman correlation coefficients across distributions and sample sizes: A tutorial using simulations and empirical data. *Psychol. Methods* **2016**, *21*, 273–290. [CrossRef]
54. Nossent, J.; Elsen, P.; Bauwens, W. Sobol' Sensitivity Analysis of a Complex Environmental Model. *Environ. Model. Softw.* **2011**, *26*, 1515–1525. [CrossRef]
55. Tian, W. A Review of Sensitivity Analysis Methods in Building Energy Analysis. *Renew. Sustain. Energy Rev.* **2013**, *20*, 411–419. [CrossRef]

56. Tortorelli, D.A.; Michaleris, P. Design Sensitivity Analysis: Overview and Review. *Inverse Probl. Eng.* **1994**, *1*, 71–105. [CrossRef]
57. Zhou, Y.; Tam, V.W.Y.; Le, K.N. Sensitivity Analysis of Design Variables in Life-Cycle Environmental Impacts of Buildings. *J. Build. Eng.* **2023**, *65*, 105749. [CrossRef]

buildings

MDPI

Article

The Effect of Low-Carbon Technology on Carbon Emissions Reduction in the Building Sector: A Case Study of Xi'an, China

Dongyi Zhang [1,†], Lu Sun [1,*,†], Yifan Zhang [1], Tianye Liu [1], Lu Gao [2], Fufu Wang [1], Xinting Qiao [1], Yuqi Liu [1], Jian Zuo [3] and Yupeng Wang [1]

1 School of Human Settlements and Civil Engineering, Xi'an Jiaotong University, Xi'an 710049, China
2 School of Economics, Management and Law, University of South China, 28 Changsheng West Road, Hengyang 421001, China
3 School of Architecture and Civil Engineering, The University of Adelaide, Adelaide, SA 5005, Australia
* Correspondence: sun.lu@xjtu.edu.cn
† These authors contributed equally to this work.

Abstract: Efficient carbon reduction pathways in the building sector are critical for urban decarbonization. This study predicts urban carbon emissions and establishes models to evaluate the carbon emission reduction potential of applying building low-carbon technologies (LCTs) at the urban scale. The models under consideration encompass a spectrum of active strategies, specifically heat pump (HP), rooftop photovoltaic (PV) systems, and smart heating, ventilation, and air conditioning (HVAC) systems, alongside passive strategies encompassing advanced building materials and building envelopes. The predictive calculations consider building typologies, technological evolution, adoption rates, and local policy constraints. Results indicate that by 2030, the building sector in Xi'an will account for over 30% of the city's total carbon emissions. The integrated emission reduction effect of LCTs reaches 25.8%, with building materials contributing the most significantly at 9%. Notably, rooftop PV systems demonstrate the highest carbon reduction potential among active strategies, while HP exhibits the fastest annual growth rate in mitigation. Furthermore, the study evaluates the feasibility of these LCTs to accelerate progress toward carbon reduction goals in the building sector.

Keywords: carbon emissions prediction; sustainable buildings; low-carbon technologies; carbon emission model; carbon reduction potential

1. Introduction

In response to the challenge of climate change, reducing greenhouse gas (GHG) emissions is an important issue in the world. China proposed achieving carbon peaking by 2030 and carbon neutrality by 2060, a goal that necessitates collaborative efforts across all sectors [1]. On a global scale, the three primary sectors responsible for substantial carbon emissions are buildings, industry, and transportation. Of these, buildings account for the highest share of energy demand, approximately 35% [2]. In China, the cumulative carbon emissions across the entire lifecycle of buildings reached 5.13 billion tons, accounting for a significant proportion of 48.3% of the total carbon emissions in 2021 [3]. A study of 28 provinces in China found that housing retrofits can reduce household energy consumption and carbon emissions by 17.87% compared to unretrofitted housing, significantly contributing to building energy efficiency and carbon reduction [4]. Given that cities serve as pivotal platforms for production and various activities, it is imperative to scrutinize the trend of carbon emissions from the building sector and assess the low-carbon potential of buildings. This analysis is crucial for the development of effective carbon emissions reduction technologies, which are indispensable in the fight against climate change, especially to achieve the goal of urban carbon neutrality [5].

To evaluate the contribution of LCTs in reducing urban carbon emissions, it is necessary to predict the future trajectories of carbon emissions within both the city and building sectors and to explore effective decarbonization pathways. For this purpose, top-down and bottom-up models are widely utilized in the development of prediction models aimed at estimating urban and building carbon emissions. Specifically, top-down models provide carbon emissions from cities and building stocks by considering factors such as carbon emissions, economics, and relevant statistics [6]. In contrast, bottom-up modeling establishes a baseline for the carbon emissions status of the assessed subject. When combined with scenario analysis, this approach can elucidate the direct impact of energy-saving and carbon-reduction measures on building energy consumption and emissions, encompassing technologies for enhancing building energy efficiency and energy substitution-related technologies. Notable bottom-up approaches include the Long-range Energy Alternatives Planning (LEAP), China Building Carbon Emission Model (CBCEM), and more. Researchers have made substantial progress in predicting building carbon emissions from a bottom-up perspective. However, this scenario analysis possesses limitations. For instance, Zhang and Luo [7] introduced a carbon emission prediction framework for public buildings in China, grounded in the LEAP model. This framework predominantly considers historical factors such as building area and energy intensity while neglecting the policy background of carbon reduction measures and the constraints of geographical natural resource conditions. Consequently, it establishes various carbon emissions scenarios for public buildings. Moreover, Tan, Lai et al. [8] utilized the bottom-up model from the Chinese Academy of Sciences to anticipate residential building carbon emissions up to 2050, including heating, urban buildings, and rural buildings. Factors such as population size, building area, urbanization rate, building energy efficiency, and changes in energy structure exert considerable influence on the overall carbon emissions during the operational phase of a building. Yuan, Chen et al. [9] found that in pursuing carbon emissions peaking within the building sector, four pivotal measures: advocating for the adoption of low-carbon and clean heating methods, extensive utilization of renewable energy sources, implementation of low-carbon retrofitting initiatives, and rational regulation of building scale. Xu, Sun et al. [10] evaluated the challenges associated with implementation, economic costs, and potential for widespread adoption of various technologies, following a rigorous calculation of carbon emissions from buildings across diverse scenarios. Their findings led to the proposal of the following prioritized strategies: firstly, enhancing the energy efficiency standards of newly constructed buildings; secondly, fostering the integration of renewable energy in building applications; and lastly, advancing energy-saving retrofits of existing buildings.

Research has demonstrated the effectiveness of clean energy and renewable energy in facilitating the decarbonization of the urban building sector. However, there is a paucity of studies focused on the selection of LCTs for achieving efficient decarbonization in buildings. The majority of existing research has primarily concentrated on the impacts of implementing LCTs on individual buildings or confined geographical areas. Consequently, the effects of integrating multiple technologies at the urban scale, and the contribution of the carbon reduction effect of different technologies to the city's carbon-peaking goals, remain in need of thorough quantification.

Existing studies have demonstrated the contributions of clean and renewable energy systems to decarbonizing urban building sectors. However, there remains a paucity of research focused on technology selection strategies for LCTs to achieve effective building decarbonization. Current research predominantly concentrates on assessing the environmental impacts of implementing LCTs at either the individual building level or within limited geographical boundaries. Consequently, two critical knowledge gaps persist: (1) quantifying the implementation efficacy of LCTs at the urban scale; (2) investigating the impact and contribution of different technologies' carbon reduction potential on urban carbon reduction. These unresolved scientific inquiries necessitate establishing systematic quantification methodologies through developing comprehensive urban-scale analytical frameworks.

This study attempts to clarify the carbon emissions reduction effect of applying LCTs in buildings, to address the following research questions: (1) What are the carbon emissions and future trends of the city and building sector? (2) What is the carbon reduction impact of typical LCT applications in urban buildings? (3) What are the effectiveness and feasibility of different LCTs in terms of reducing carbon emissions?

This study fills the research gaps in two aspects: Firstly, it quantifies the carbon reduction potential of pivotal LCTs applied in the target city, expanding the research scope from small-scale building clusters to the urban scale; Secondly, it conducts comprehensive quantitative analysis of the contribution of LCTs and carbon sinks towards achieving urban carbon peaking goals.

2. Materials and Methods

The framework of this study is shown in Figure 1. The first step is to construct the database, including the energy consumption of various industries and building-related data of the city, etc. The LEAP model is widely used in the medium and long-term national and city energy and environmental planning, and it is capable of forecasting the long-term energy supply and demand dynamics of society under the influence of various drivers. In addition, the model can estimate the pollution and greenhouse gas emissions generated by the energy distribution and consumption processes [11]. Utilizing the LEAP model, take into account local policies and objective conditions, and establish the scenario parameters of low-carbon measures under different intensities, to predict the multi-scenario carbon emission results that conform to the actual situation of the city. Secondly, select five typical LCTs in the city, considering the building types, technological iterations, renovations, and application rates in the city over time (year), and establish corresponding models to calculate the carbon emission reduction amounts resulting from the application. Thirdly, to conduct a comparative analysis of the effects of these technologies on urban carbon emissions reduction. Finally, examine the feasibility of technology applications and put forward suggestions for the city to achieve a low-carbon peaking target.

2.1. Research Area

This study takes Xi'an City, located in Shaanxi Province, as a case study. The operational carbon emissions of the construction industry in Shaanxi Province witnessed a substantial surge, rising from 0.5 billion tons in 2010 to 0.8 billion tons in 2020, marking a growth rate of 60%. Notably, the amount of new construction and Retrofittable in Xi'an has increased significantly in recent years. By the end of 2023, Xi'an's resident population exceeded 13 million, accounting for more than 30% of the province's population, and had been increasing except for a slight negative growth in 2021. Furthermore, Xi'an's Gross Domestic Product (GDP) exhibited robust performance, reaching 1,201,076 billion yuan in 2023, maintaining high growth. During the period from 2016 to 2020, Xi'an achieved cumulative reductions of approximately 24% in energy consumption per unit of GDP and 19% in carbon emission intensity, exceeding the provincial target.

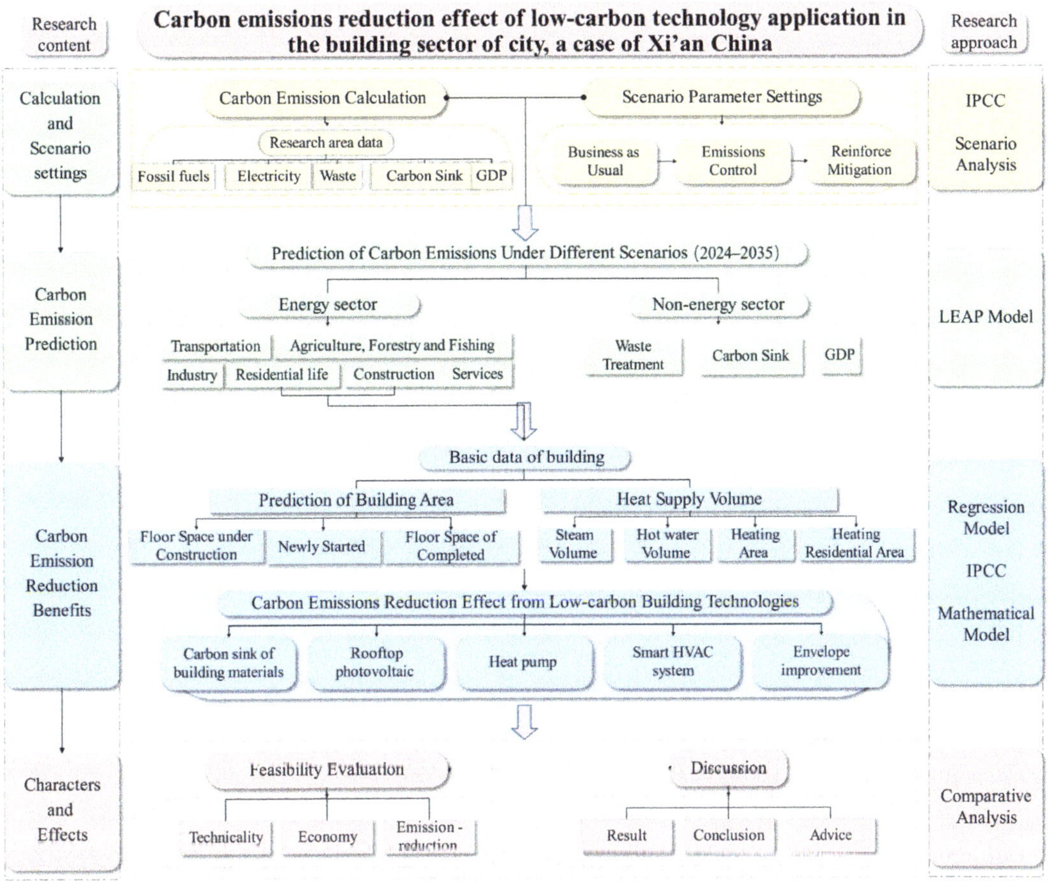

Figure 1. Research framework.

2.2. *Carbon Emission Analysis of Xi'an*

Calculation of the Carbon Emissions

The energy consumption data were obtained from the official report of the Urban Bureau of Statistics, as shown in Tables S1 and S2. The total carbon emissions from energy consumption were derived from aggregated activity-level data. The formula is as follows:

$$CE_{ec} = \sum EC_{i,j} \times c_i \tag{1}$$

where CE_{ec} represents the carbon emissions from net energy consumption, $EC_{i,j}$ is the net consumption of energy i in industry sector j, and c_i is carbon emission factor of energy i. The same principle applies to solid waste treatment.

Methane emissions generated from urban wastewater treatment can be converted into carbon dioxide equivalent using Global Warming Potential (GWP) values. The formula is as follows.

$$CE_{wt} = Q \times \Delta COD \times c_{CH_4} \times GWP_{CH_4} \tag{2}$$

where CE_{wt} represents the carbon emissions from the treatment of wastewater, Q is the annual treated wastewater volume, and c_{CH_4} is the methane emission factor, and GWP_{CH_4} is the GWP of methane.

The green space carbon sink calculations reference the "Technical Guidelines for Gross Ecosystem Product (GEP) Accounting of Terrestrial Ecosystems" [12].

2.3. Prediction of the Future Carbon Emissions Under Different Scenarios

The LEAP model serves as an energy system planning model that considers a variety of aspects. In this study, the LEAP model was employed to simulate future carbon emission scenarios, methodologically grounded in the aggregation of multi-sectoral energy consumption data and the incorporation of policy constraints (Figure 2). This approach enables the evaluation of the impacts of diverse low-carbon interventions on urban carbon peaking timelines and peak magnitudes.

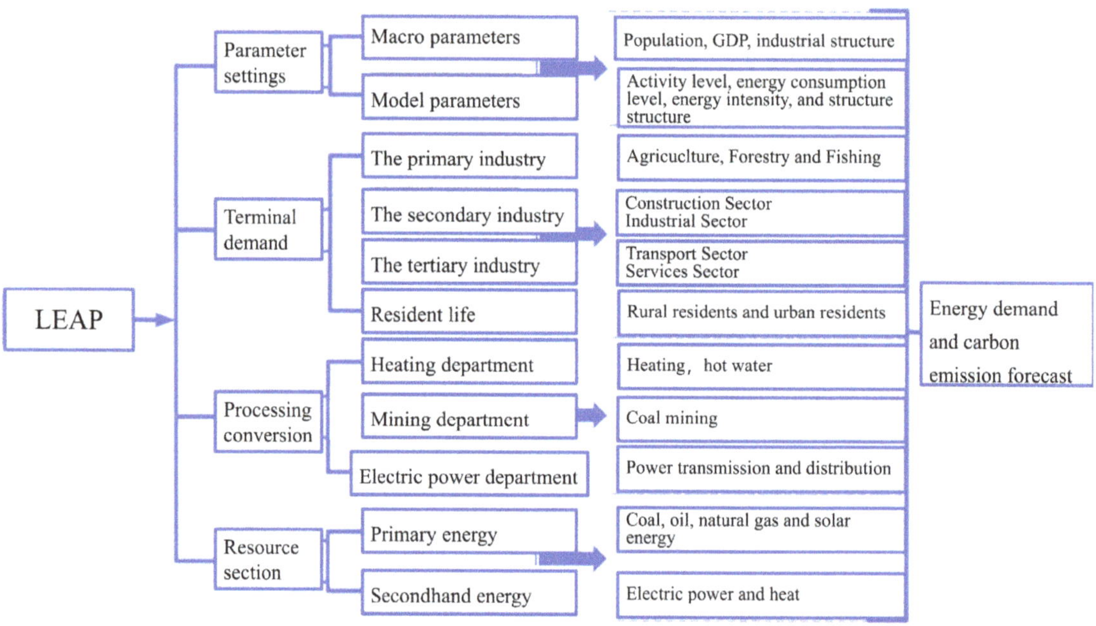

Figure 2. LEAP model framework.

The scenario modeling incorporated three different scenarios: the business as usual (BAU) scenario, the Emission Control (EMC) scenario, and the Reinforced Emission Reduction (RFM) scenario. The BAU scenario outlines the anticipated carbon emissions trajectory in the future under the assumption of no policy interventions, characterized by the fastest GDP growth based on current policies. In contrast, the EMC scenario appropriately raises the share of clean energy and reduces the proportion of fossil energy, resulting in moderate GDP growth. Lastly, the RFM scenario achieves the most significant energy savings and emission reductions but is likely accompanied by a decline in the GDP growth rate.

The constant growth rate assumption was employed in the modeling of the scenario setting, given the availability of data, the accuracy of the data, the study objectives, and the regional characteristics. This assumption was made in order to simplify the modeling. In establishing the scenario parameters for this study, the process of "input-output-judgment result-iterative input-output" was repeated multiple times to ensure that the overall forecasting results would not be significantly influenced by the variation in a specific parameter.

The reference basis for the model parameter settings is shown in Table S3, and the parameter settings are presented in Table 1.

2.4. Carbon Emissions Reduction Effect from Building Low-Carbon Technologies

2.4.1. Prediction of Construction-Related Data

(1) The built area prediction in the city

Table 1. Simulation parameter settings of different scenarios.

Scenario	Classification	2025	2030	2035
Business as usual (BAU)	Fossil fuels	1. Excluding the transportation industry, fossil fuels such as crude oil, gasoline, diesel, and liquefied petroleum gas in other sectors will decrease by 0.8% annually. 2. In transportation, gasoline and diesel consumption decrease by 5% annually. 3. The consumption of natural gas in all sectors rises by 4% annually.		
	Electricity	The consumption of electricity power in all sectors rises by 6.6% annually.		
	Solid waste and sewage treatment	The annual rate of reduction for both solid waste generation and the Chemical Oxygen Demand (COD) in wastewater is 2%.		
	Clean energy	The proportion of clean energy generation has reached 20%.	50%.	70%.
	Forestry carbon sink	The area covered by grassland and forest has been expanded to encompass 48.03%	49%	50%
	GDP	The annual growth rate of the GDP is 6.5%.		
Emissions control (EMC)	Fossil fuels	1. Excluding the transportation industry, fossil fuels such as crude oil, gasoline, diesel, and liquefied petroleum gas in other sectors will decrease by 1.5% annually. 2. In the transportation sector, the consumption of gasoline and diesel decreases by 7.5% per year. 3. The consumption of natural gas in all sectors rises by 6% annually.		
	Electricity	The electricity consumption of electricity power in all sectors rises by 10% annually.	10.5% annually.	11% annually.
	Solid waste and sewage treatment	The annual rate of reduction for both solid waste generation and the Chemical Oxygen Demand (COD) in wastewater is 2%.	3%.	4%.
	Clean energy	The proportion of clean energy generation has reached 25%.	55%.	75%.
	Forestry carbon sink	The area covered by grassland and forest has been expanded to encompass 50%	52%	54%
	GDP	The annual growth rate of the GDP is 6.5%.	6%.	5.5%.

Table 1. *Cont.*

Scenario	Classification	2025	2030	2035
Reinforce mitigation (RFM)	Fossil fuels	1. Excluding the transportation industry, fossil fuels such as crude oil, gasoline, diesel, and liquefied petroleum gas in other sectors will decrease by 3% annually. 2. In the transportation sector, gasoline and diesel consumption decrease by 10% per year. 3. The consumption of natural gas in all sectors rises by 7% annually.		
	Electricity	The electricity consumption of electricity power in all sectors rises by 11% annually.	11.5% annually.	12% annually.
	Solid waste and sewage treatment	The annual rate of reduction for both solid waste generation and the Chemical Oxygen Demand (COD) in wastewater is 4%.	5%.	6%.
	Clean energy	The proportion of clean energy generation has reached 25%.	60%.	80%.
	Forestry carbon sink	The area covered by grassland and forest has been expanded to encompass 50%	53%	56%
	GDP	The annual growth rate of the GDP is 6%.	5%.	4%.

The newly developed and completed built-up areas within Xi'an from 2013 to 2022 were sourced from the Xi'an Bureau of Statistics (Table S4). Linear regression and curve regression models are employed to predict the built-up area, newly constructed area, and completed built-up area, respectively. The *p*-value and R^2 results showed that the prediction models are verified to be used [13]. The expressions are shown in Table 2.

Table 2. Prediction results of the built area data of Xi'an.

Category	Expression	*p* Value	Fitting Degree R^2	F Test
Built-up area (10,000 m^2)	$S_1 = 1685.053 \times N - 3{,}383{,}658.772$	0.000	0.943	F = 132.845 (p = 0.000 ***)
New built area (10,000 m^2)	$S_2 = 291.313 \times N - 583{,}264.641$	0.001	0.851	F = 24.154 (p = 0.001 **)
Completed built area (10,000 m^2)	$S_3 = 123.314 \times N - 246{,}004.91$	0.002	0.819	F = 20.455 (p = 0.002 *)

Note: N is the year. *: $p < 0.05$ ("significant", moderate evidence for rejection of the original hypothesis.). **: $p < 0.01$ ("highly significant", strong evidence for rejection of the original hypothesis.). ***: $p < 0.001$ ("highly significant", very strong evidence for rejecting the original hypothesis.).

(2) The heating demand prediction

The annual steam and hot water demand data, heating area, and residential area of the city from 2013 to 2022 were selected from the Xi'an Bureau of Statistics (Table S5). The average value was adopted as the predicted value of the total steam volume, while the linear regression model was utilized for the remaining variables. The expressions are shown in Table 3.

Table 3. Prediction results of the heating demand of Xi'an.

Category	Expression	p-Value	Fitting Degree R^2
Total steam volume (10,000 GJ)	$Q_1 = 1843.927$	-	-
Total hot water volume (10,000 GJ)	$Q_2 = 661.782 \times N - 1{,}329{,}014.105$	0.000 ***	0.920
Heating area (10,000 m^2)	$S_4 = 2603.1 \times N - 5{,}228{,}494.216$	0.000 ***	0.887
Residential heating area (10,000 m^2)	$S_5 = 1905.564 \times N - 3{,}825{,}721.505$	0.001 **	0.846

Note: N is the year. **: $p < 0.05$ ("significant", moderate evidence for rejection of the original hypothesis.). ***: $p < 0.01$ ("highly significant", strong evidence for rejection of the original hypothesis.).

2.4.2. Carbon Emissions Reduction Effect from Building Materials Carbon Sink

Building materials carbon sink encompasses the active role that buildings and associated activities play in the carbon cycle, involving the absorption, storage, and reduction of carbon dioxide emissions [14]. Zhao, Huang et al. [15] formulated a generalized model to predict the materials' carbon sink of urban buildings across vast areas. Drawing Fick's second law, it is established that the carbonation depth scales proportionally with $t^{0.5}$. Employing Pearson correlation analysis [16,17], among the eight independent variables, the effective exposed area A_e (with the best correlation) was incorporated as another parameter of the prediction model. The power function regression model with the optimal goodness of fit R^2 was utilized to predict the building carbon sink in Xi'an. The formula is as follows:

$$CER_{bm} = CER_{bm,RB} + CER_{bm,IB} + CER_{bm,OB} + CER_{bm,CB} + CER_{bm,PB} \quad (3)$$

$$CER_{bm,i} = a_1 \times A_{i,t}^{a_2} \times t^{0.5} \quad (4)$$

where CER_{bm} represents the carbon emission reduction from building materials carbon sink, $A_{i,t}$ is the exposed area of building type i with a completion age of t years (Table S6), RB is residential buildings, IB is industrial buildings, CB is commercial buildings, PB is public service buildings, OB is other buildings, a_1 and a_2 are known constant coefficients

Given the correlation between the building materials' carbon sink and their completion time, this study focuses specifically on the carbon sinks associated with buildings that were completed in 2013 and thereafter.

2.4.3. Carbon Emissions Reduction Effect from Building Heat Pumps

The building sector consumes a significant amount of heat, with the heat source structure mainly coal-fired [18]. As a green and low-carbon heat supply scheme, heat pumps (HPs) are essential for the building sector to replace fossil energy for heating and achieve carbon-peaking goals and neutrality. The carbon emissions reduction effect from HP is the difference between the carbon emissions in the Baseline Scenario (BS) and the carbon emissions from using HP for heating in the Application Scenario (AS). The formula is as follows:

$$CER_{hp} = CER_{hp,Q_1} + CER_{hp,Q_2} \quad (5)$$

$$CER_{hp,i} = CE_{hp,BS} - CE_{hp,AS} \quad (6)$$

$$CE_{hp,BS} = \frac{\Sigma Q_j}{q_i} \times c_i \quad (7)$$

$$CE_{hp,AS} = \frac{\Sigma Q_j \times TAP_i / HSFP_k}{q_i} \times c_i + (1 - TAP_i) \times CE_{hp,\ BS} \quad (8)$$

where CER_{hp} represents the carbon emission reduction from HP, Q_1 is the total annual steam demand, Q_2 is the total annual hot water demand, q_i is the average calorific value of energy i, c_i is the carbon emission factor of energy i, *HSFP* is the Heating Season Performance Coefficient of type k, and *TAP* is the Technology Application Proportion.

The HSPF of different types of HP are as follows: for heating, the HSPF of household HP is 2.8, and that of commercial HP is 3; for hot water, the HSPF of household HP is 2.4, and that of commercial HP is 2.5 [19]. The baseline scenario settings are shown in Table 4 and is applied from the China Energy Conservation Association's "Heat Pump Development Roadmap for China's Building Sector".

Table 4. Baseline scenario settings for heating and hot water supply.

Heat Pump Market Type		Baseline Scenario	Main Applicable Scenarios
Heating	Household heat pump heating unit	100% coal burning	for residential heating in severe cold and cold regions
		100% gas wall-hung boiler	for residential heating in hot summer and cold winter regions
	Heat pump air heater	100% coal burning	for residential heating in severe cold and cold regions
	Commercial heat pump unit	50% cogeneration + 10% gas boiler + 40% coal-fired boiler	for central heating in severe cold and cold regions
	Water source heat pump unit		
Hot water	Household heat pump water heater	50% gas water heater + 50% electric water heater	for the hot water supply in all regions
	Commercial heat pump water heater		

The market growth rate of HP is estimated using nodal sales data for hot water and heating applications under carbon neutrality scenarios (Table S7). The curvilinear regression model with a goodness-of-fit of 0.996 is selected as the predictive model for simulated sales, with it is shown in Table 5.

Table 5. Prediction model for China's heat pump sales.

Type	Expression	*p* Value	Fit Degree R^2
Hot water sales	$Q_3 = 0.334 \times N^2 - 1308.235 \times N + 1{,}282{,}101.84$	0.000 ***	0.996
Heating sales	$Q_4 = 1.368 \times N^2 - 5509.464 \times N + 5{,}548{,}091.723$	0.000 ***	0.996

Note: N is the year. ***: $p < 0.001$ ("highly significant", very strong evidence for rejecting the original hypothesis.).

The growth in HP market sales typically directly reflects the prevalence of technology adoption; therefore, its annual growth rate is utilized as a proxy for the *TAP* (Table S8).

2.4.4. Carbon Emissions Reduction Effect from Rooftop Photovoltaic Systems

Given the vigorous development of the urban economy, the high concentration of power demand, and the abundant building roof resources [20], roof PV systems not only refrain from interfering with the original functions of the building but also have the potential to achieve self-sufficiency in power supply [21]. The formula is as follows:

$$CER_{pv} = \Sigma(RUC \times RA_i) \times SI \times CEM \times OE \times c \quad (9)$$

where CER_{pv} represents the carbon emission reduction from PV, RUC_i is the Roof Utilization Coefficient, RA_i is the roof area of buildings, i is the building category, SI is the total solar irradiance received by PV systems, CEM is the conversion efficiency of PV modules, and OE is the PV operational efficiency of, c is the carbon emission factor of electricity.

The RUC is 0.4 for residential land, 0.6 for public and commercial land, and 0.8 for industrial land [22]. The average total horizontal irradiation in Xi'an in 2022 is 1459.7 kWh/m^2. The conversion efficiency of PV modules in 2024 is assumed to be 16.7%, with an annual growth rate set at 0.7% [23,24]. The operational efficiency of PV is defined as 0.8 [25]. An average tilt angle of 20° is assumed in the roof PV system model. This angle is corrected by considering the discrepancy between the theoretical optimal tilt angle (34°) and the error caused by assuming 0°, as well as practical limitations such as roof load-bearing capacity, esthetic requirements, multi-row installation spacing constraints, and the need to balance power generation efficiency with high-temperature attenuation, to ensure both theoretical and engineering feasibility.

2.4.5. Carbon Emissions Reduction Effect from Smart Heating, Ventilation, and Air Conditioning Systems

There is a significant demand for air conditioning in commercial buildings, office buildings, and large shopping malls [26]. Smart heating, ventilation, and air conditioning (HVAC) systems dynamically regulate indoor environmental parameters (e.g., temperature, humidity, and airflow), saving energy while maintaining thermal comfort [27]. The formula is as follows:

$$CER_{hvac} = \Sigma EC_{c,i} \times A_i \times ES_i \times c \tag{10}$$

where CER_{hvac} represents the carbon emission reduction from HVAC, $EC_{c,i}$ is the unit energy consumption for building cooling, A_i is the area of building type i, ES_i is the energy-saving rate, and c is the carbon emission factor of electricity.

The annual cooling energy consumption is quantified as 8 kWh/(m^2·a) for RB, 37 kWh/(m^2·a) for CB, 17 kWh/(m^2·a) for PB, and 20 kWh/(m^2·a) for other building typologies. The current energy-saving rate is set at 20%, while mandatory energy performance standards mandate a 50% improvement by 2035 relative to 2023 technological baselines. Based on these constraints, the annual energy efficiency improvement rate is calibrated. The analysis encompasses buildings completed in or after 2013 [28].

2.4.6. Carbon Emissions Reduction Effect from Building Envelope Improvement

Common technologies for building envelope improvement include external wall insulation systems, reflective coatings, and thermal insulation coatings [29]. The application of high-performance materials enhances building thermal performance primarily through the integration of materials with superior thermal insulation, soundproofing, fire resistance, and other properties [30]. Representative materials encompass aerogels, vacuum insulation panels (VIPs), and phase change materials (PCMs) [31]. This study quantifies the carbon emission reduction benefits of upgrading to more efficient insulation materials using the following formula:

$$CER_{ei} = \Delta HL \times \sum {}^{q_i}/_{c_i} \tag{11}$$

$$\Delta HL = \Delta U \times HFA \times \Delta T \times \sum q_i \times c_i \tag{12}$$

where CER_{ei} represents the carbon emission reduction from envelope improvement, ΔHL is the reduced heat loss amount, U is the thermal transmittance, HFA is the heated floor area. ΔT is the average indoor/outdoor temperature difference during winter heating in Xi'an. q_i is the average calorific value of energy i, and c_i is the carbon emission factor of energy i.

Based on the national standards for thermal performance benchmarks of building envelopes [32–34], Xi'an is classified as Cold Region Zone B, with the reference heat transfer coefficient (K_0) of its envelope structures set at 0.35 W/(m^2·K) [35]. The technology is assumed to achieve approximately 15% load reduction. The average indoor/outdoor temperature difference during the heating season in Xi'an is defined as 20 °C, with a heating duration of approximately four months.

3. Results

3.1. Carbon Emissions Characters of Xi'an

The total carbon emissions comprise the following sectors: industry (1.65×10^7 t), service industry (1.20×10^7 t), transportation industry (8.31×10^6 t), residential life (1.352×10^7 t), construction industry (3.95×10^6 t), and agriculture, forestry, animal husbandry, and fishery (7.09×10^5 t). Among these, the industrial sector contributes a large proportion, accounting for 30.60% of total emissions, followed by residential living (25.07%), the service industry (22.19%), transportation (15.41%), construction (7.33%), and agriculture, forestry, animal husbandry, and fishery (1.31%).

The predicted results of carbon emissions in Xi'an under different scenarios are shown in Figure 3. Under the BAU scenario, Xi'an is projected to achieve its carbon peak target by 2029, with total carbon emissions reaching 5.62×10^7 tons (Table S9). In contrast, the EMC scenario demonstrates accelerated progress, enabling the city to attain carbon peaking one year earlier in 2028, with corresponding emissions reduced to 5.45×10^7 tons (Table S10)—representing a 3.02% reduction compared to the BAU baseline. Under the RFM scenario, through intensified energy conservation and emission reduction measures, Xi'an could advance its carbon peak timeline to 2026. This pathway would result in peak emissions of approximately 5.4×10^7 tons (Table S11), corresponding to a 3.20% decrease relative to the BAU scenario.

3.2. Carbon Emissions Reduction Effect from Building Low-Carbon Technologies of Xi'an

As illustrated in Figure 4, the implementation of various low-carbon building technologies in Xi'an City has yielded a demonstrable impact on carbon reduction.

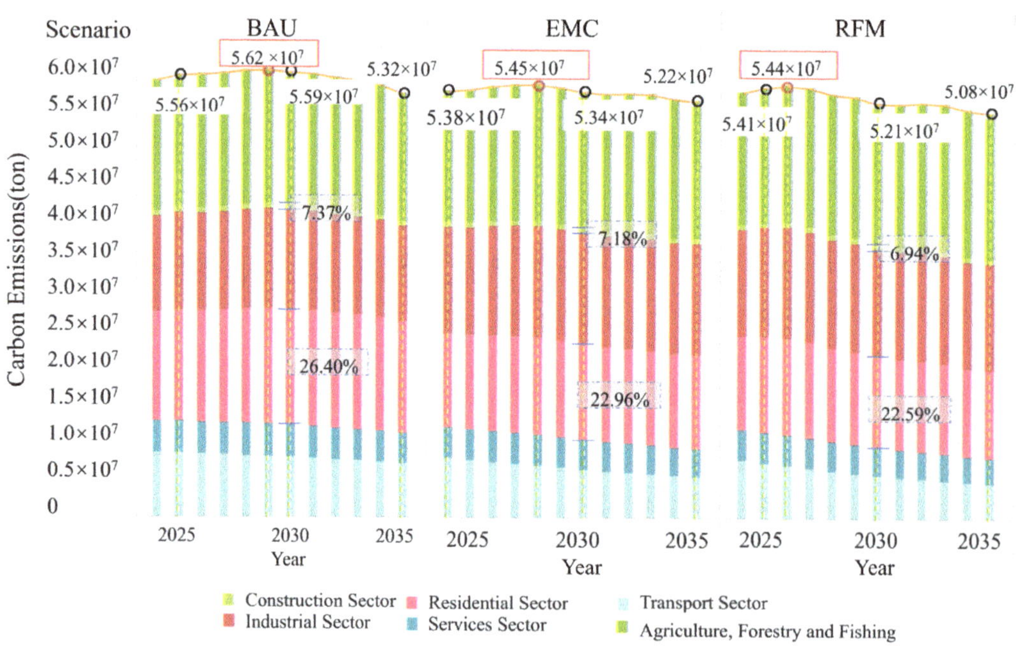

Figure 3. Multi-scenario projections of carbon emissions of Xi'an.

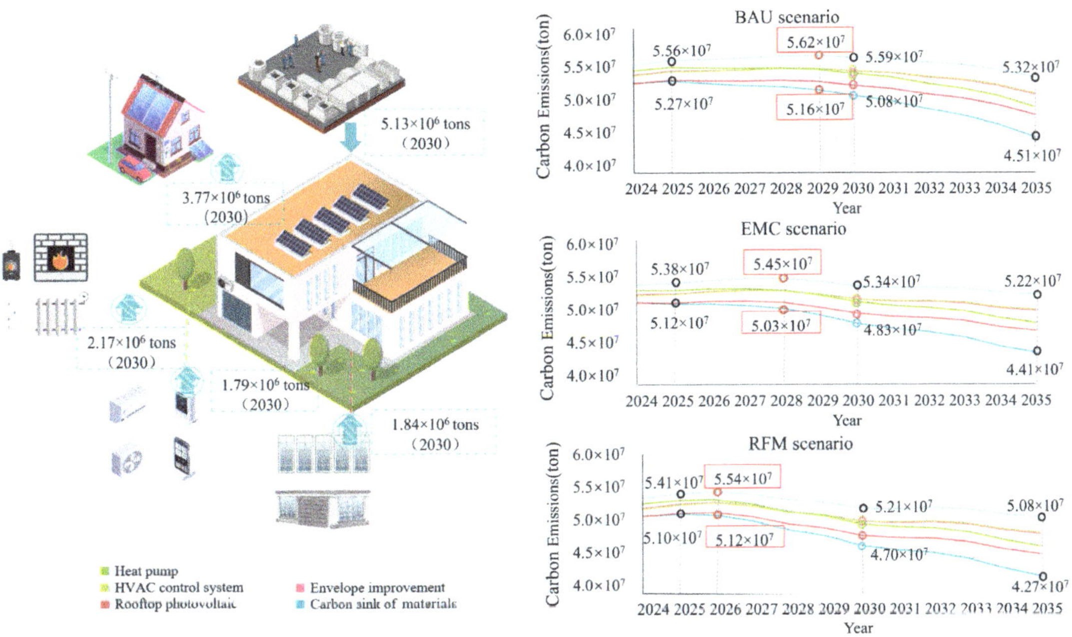

Figure 4. Carbon emissions reduction effect of LCT applications.

The building carbon sinks continue to accumulate in the research period (Table S12). By 2035, it is estimated that the reduction in carbon emissions due to building carbon sinks will reach 8,095,794 tons. Among these reductions, the contributions from residential buildings (RB), public service buildings (PB), and other buildings (OB) are relatively significant, accounting for 22.98%, 25.75%, and 31.59%, respectively. In 2024, building carbon sinks accounted for over 4% of Xi'an's total carbon emissions. This proportion is projected to increase to approximately 9% by 2030.

With the expansion of the scale of building heating and hot water using HP, it is predicted that by 2035, the annual carbon emissions reduction contribution of HP in Xi'an will exceed 40% of the carbon emissions associated with heating and hot water in the baseline scenario (Tables S13 and S14). In 2024, the carbon emissions reduction achieved through the application of building HP accounted for approximately 1% of Xi'an's total carbon emissions. By 2030, this proportion is anticipated to increase to around 4%.

As shown in the results (Table S15), the carbon emission reductions achieved through solar power generation by rooftop PV systems are projected to reach 3.5 × 10^6 tons by 2030. In 2024, the carbon reduction from rooftop PV systems accounted for approximately 4% of Xi'an's total carbon emissions. This proportion is projected to increase to around 6% by 2030.

The carbon emissions reduction from smart HVAC systems can reach 1.79 × 10^6 tons in 2030 (Table S16). In 2024, the carbon emissions reduction attributed to the smart application of smart HVAC systems in buildings accounts for approximately 2.7% of the total carbon emissions in Xi'an. By 2030, this proportion will rise to around 3.3%.

By 2030, the reduction in heat loss achieved through envelope insulation improvements is quantified at 1.41 × 10^3 GJ (Table S17). In 2024, carbon emission reductions attributed to enhanced thermal performance of building envelopes accounted for approximately 2.5% of Xi'an City's total carbon emissions. This proportion is projected to rise to around 3.5% by 2030.

From the perspective of the carbon reduction potential of technology applications at the urban scale (Figure 5), in 2024, the carbon sink of building materials accounts for the highest proportion (approximately 30%) of total carbon reduction from all technologies. This is followed by rooftop PV systems (28%), smart HVAC systems (18%), envelope improvements (16%), and HP (~9%). In terms of growth rates, by 2030, building carbon sinks will remain the primary contributor to urban carbon reduction, representing approximately 35% of the total. The proportion of carbon reduction from HP rises to third place (15%), with its growth rate consistently ranking highest.

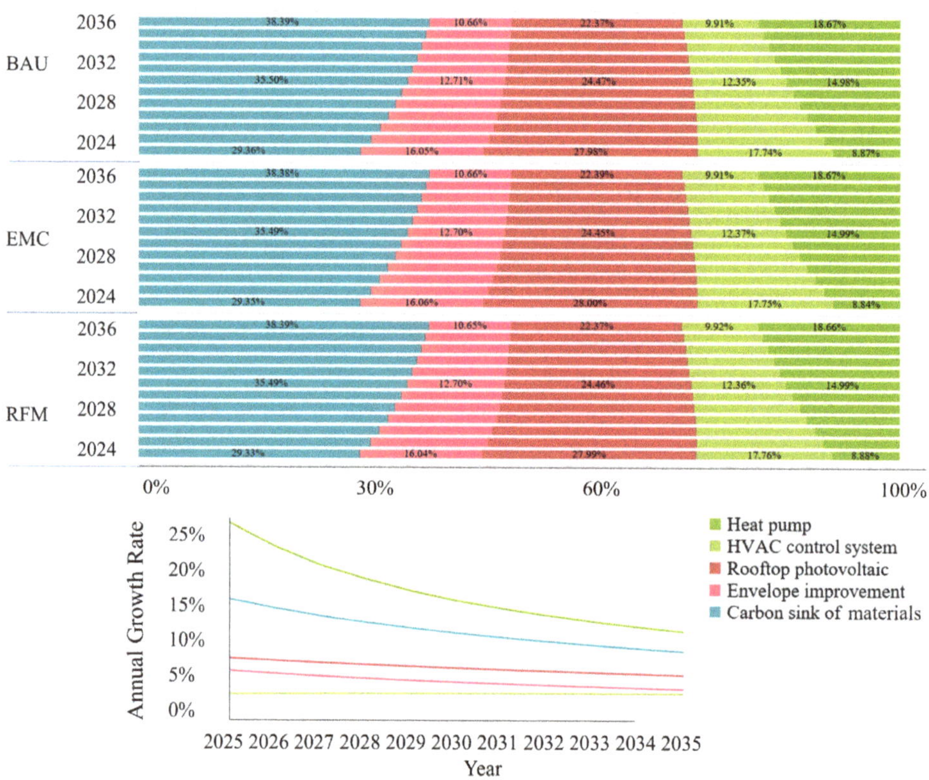

Figure 5. Contribution comparison of building LCTs.

4. Discussion

4.1. The Urban Building Sector's Substantial Carbon Mitigation Potential

Under various scenarios, carbon emissions from building operations exceed 30% of total urban emissions. Across multiple scenarios, the urban building sector consistently demonstrates the highest emission reduction potential, which aligns with previous research findings [36]. Our findings demonstrate that implementing LCTs in Xi'an's building sector can reduce CO2 emissions by 43.3% within the sector and 14.2% citywide by 2024, projected to rise to 76.3% and 25.8%, respectively, by 2030. Operational phases account for the largest share of building-related emissions, with LCTs capable of halving these emissions. Comparative analysis reveals that the reinforced mitigation scenario not only accelerates the carbon peaking timeline but also effectively reduces the magnitude of the emission peak, thereby aligning with economic development objectives and establishing a more sustainable pathway toward long-term climate goals.

The study comprehensively assessed the effects of multiple types of technologies, rather than a single measure, on the subject. This outcome serves to illustrate the combined impact of low-carbon technologies. The urbanization process is staged, and the projection year is situated in the middle stage of urbanization. The reduction in energy consumption is achieved through energy efficiency retrofits in existing buildings and the implementation of higher energy efficiency standards in new buildings. Collectively, these factors contribute to the decline in the sectoral share of emissions.

4.2. Differences in Emission Reduction Potential and Feasibility of Low-Carbon Technologies

Among the contributions of different LCTs, the carbon sink of building materials stands out. Green building materials, distinguished by their inherent advantages, have emerged as a pivotal strategy for the building materials industry to attain carbon reduction objectives [37]. The decarbonization potential of building materials is particularly noteworthy, given their inherent capacity for carbon sequestration. The production of building materials (e.g., cement, steel, glass) is known to be a significant consumer of energy and a substantial source of emissions. It has been estimated that the carbon emissions associated with the production of these materials account for more than 50% of the total carbon emissions emitted during the lifecycle of a building. However, when utilized in architectural settings, these materials possess the capacity to persist in reacting with airborne substances, thereby engendering an absorption effect. As a rapidly urbanizing area, Xi'an has a substantial building stock and a large scale of new construction, with a significant demand for materials. Consequently, there is considerable scope for optimizing material choices (e.g., promoting low-carbon concrete and recycled building materials) to reduce emissions. Given the scarcity of land resources, China and the majority of Asian countries share similarities in the structural composition of residential and commercial buildings, with reinforced concrete structures being prevalent. Conversely, in regions such as North America and Australia, wood structures are more commonly adopted for building construction [38].

Traditional energy sources such as coal, used for heating, power generation, and cooling, impose significant environmental burdens. Enhancing energy efficiency and increasing the adoption of renewable energy, particularly solar power, can mitigate these impacts by reducing emissions and conserving resources. Among the evaluated retrofitting solutions, the application of rooftop PV systems demonstrates a remarkable increase in clean energy integration, highlighting its exceptional potential for lowering carbon intensity. HP exhibits the highest annual growth rate in carbon reduction efficiency, underscoring its promising applicability in energy transition scenarios. Additionally, smart HVAC system upgrades and building envelope retrofits contribute significantly to reducing reliance on conventional energy sources. Notably, solar energy harnessed through PV systems not only supplies clean power for building operations but also enables synergies with other LCTs, such as energy storage and smart grid integration.

Among active LCTs, rooftop PV systems in Xi'an demonstrate the most significant carbon emission reduction, underscoring the low-carbon benefits of renewable energy systems. This phenomenon can be attributed to the synergistic effect of its climatic conditions, application scenarios, and policy strength. Xi'an experiences an annual mean of 2000 to 2500 h of sunshine, which is sufficient to support the basic performance of rooftop PV systems. The extensive roof area offers significant potential for integrating PV technology into the urban infrastructure and people's lives. The city of Xi'an has clearly articulated a policy of "increasing the installed capacity of renewable energy to 30% by 2025" and has placed a significant emphasis on enhancing the proportion of PV energy. This commitment is further supported by the promotion of large-scale grid-supporting initiatives and financial subsidies to ensure that the city's extensive rooftop PV potential is fully developed.

The HP system demonstrates a high degree of scalability. The energy efficiency ratio (COP) of the heat pump is notably elevated in comparison with conventional electric heating systems or coal-fired boilers. Consequently, the carbon emission intensity per unit of energy consumption is substantially mitigated. The heat pump has been demonstrated to be a viable solution for meeting Xi'an's winter heating and hot water demands. Its suitability extends to both new construction and existing building renovation. Concurrently, the heat pump system demonstrates compatibility with renewable energy sources, such as photovoltaic power drives. This aligns with policy directives promoting the transition to electric heating and the pursuit of clean heating solutions. In the context of the northern clean heating transformation, the system benefits from financial support and market recognition.

The effect and growth rate of HVAC (heating, ventilation, and air conditioning) systems and envelope improvements in reducing emissions in the building sector are relatively unremarkable and are mainly constrained by factors such as technology, cost, implementation difficulty, and policy. At the technical level, HVAC system renovation is easily affected by the original layout of the building's pipeline network, equipment compatibility, and load demand fluctuations. Part of renovating old buildings needs to be synchronized with upgrading infrastructure. The comprehensive benefits of releasing the cycle are longer. Although envelope improvement can reduce heat loss, the energy-saving effect is more evident in the long-term reduction in energy consumption. The short-term reduction of carbon emissions is less than one would intuitively expect from renewable energy alternative technology. In terms of implementation and cost, the initial investment for HVAC retrofitting is high due to the professional and technical complexity. Building owners lack motivation to transform due to the long cost recovery period. Envelope improvement involves the interests of many parties and is vulnerable to property rights division, construction nuisance, and other constraints when transforming existing buildings. Additionally, compared to the clear subsidies and installation targets of PV projects, the energy-saving retrofit support policy for HVAC and envelope structures lacks quantitative assessments and differentiated subsidies. This makes it difficult for market players to participate in the initiative and grow rapidly. However, as technology improves and the carbon trading market improves, the energy-saving potential and growth rate of the two are expected to increase.

A triad of factors undergirds regional disparities in technology implementation: climatic conditions predominate in determining technological aptness; the economic development level and energy structure are salient; and local policies and infrastructure are also significant. Local governments' carbon peaking action programs, which delineate technology pathways, will directly impact the deployment pace.

While emission growth rates in the building sector show declining trends, achieving carbon peaking by 2030 requires further intervention to offset residual emissions. Xi'an's decarbonization strategy must prioritize technological retrofits in the building sector. Data-driven predictive modeling should guide policy design to accelerate emission reduction trajectories and align with urban carbon peaking targets. Regionally tailored low-carbon pathways, informed by localized data analytics, offer replicable frameworks for national and global scalability. Strengthening intercity and international collaboration—particularly in resource sharing and policy innovation—is imperative to address climate change collectively.

4.3. Limitations and Prospects

Given the variations in carbon emission reduction policies and targets across different regions, this paper focuses on Xi'an as a case study and utilizes precise data spanning from 2013 onwards for future predictions and calculations. Initially, the insufficiency of historical data probably has a certain impact on the prediction accuracy. Furthermore, in evaluations of the carbon reduction benefits associated with LCTs in buildings, the calculation method for each building can be further refined and improved.

The calculation framework and results of this paper can provide valuable insights and serve as a reference for cities similar to Xi'an. Furthermore, they can assist cities in different regions to calculate the corresponding carbon emission reduction pathways within the building sector. In the future, this paper intends to collect more data to conduct more accurate predictions. By further categorizing building types, the calculations will be refined to enhance the accuracy of the results. Additionally, future research should select different types of cities for calculation and compare their results to explore the differences in low-carbon pathways for buildings in different cities. Concurrently, the expanding model database will assist city authorities in planning efficiently.

5. Conclusions

The building sector, responsible for over 30% of urban carbon emissions, holds critical accountability in mitigating its environmental footprint due to its substantial energy consumption. This study establishes a comprehensive framework to quantify emission reductions from LCTs in Xi'an, integrating the carbon sequestration potential of building materials, HP systems, rooftop PV systems, smart HVAC systems, and envelope improvements. Our city-scale simulation model projects emission trajectories from 2024 to 2035 under existing policy conditions, evaluating technological applicability, regional feasibility, and iterative innovation impacts. The framework integrates the characteristics of the urban environment and infrastructure in the modeling process, ensuring the universality of the calculation indicators and emphasizing the flexibility of local adaptation. The parameters are based on localized data, with scenarios based on local policies, solar radiation based on measured values, heat pump scenarios based on climate zoning codes, and key parameters of HVAC and envelope precisely matching regional climate conditions. The adaptability of the framework across regions, particularly in small and medium-sized cities with limited data resources or unique climate conditions, overcomes the constraints of traditional models.

The principal findings demonstrate that the building sector, as a primary urban carbon emission source, can achieve carbon peaking before 2030. Quantitative assessments identify building materials in Xi'an as having the highest decarbonization potential, underscoring the strategic imperative of embodied carbon management. Within active LCTs, rooftop PV systems demonstrate optimal carbon emission reduction effect, highlighting the low-carbon advantages of renewable energy systems. HP systems exhibit rapid scalability in both retrofitting and new construction applications, demonstrating significant emission reductions through energy efficiency gains. Regional applicability analysis further reveals technology-specific deployment potentials across geographic contexts. Under business as usual (BAU) scenarios, the sector's contribution to urban emissions declines from 32.7% (2024) to 29.7% (2030), evidencing substantial mitigation effectiveness from implemented LCTs.

The synergistic deployment of these technologies not only reduces emission growth rates but also accelerates peaking timelines through amplified mitigation effects. Energy structures progressively transition toward cleaner alternatives, enhancing environmental and economic co-benefits. This research underscores the necessity for city-specific quantitative frameworks to prioritize retrofit measures across diverse climatic and socioeconomic contexts, offering actionable policy insights for data-driven low-carbon transitions. The methodology demonstrates replicable value in guiding regional climate cooperation and resource-sharing strategies to achieve urban decarbonization goals.

Supplementary Materials: The following supporting information can be downloaded at https://www.mdpi.com/article/10.3390/buildings15121989/s1. Table S1: Fundamental Metrics and Core Data of Xi'an; Table S2: Energy Consumptions by Sector of Xi'an; Table S3: Policy Objectives Reference for Scenario Settings; Table S4: Floor Space of Buildings under Construction & Completed by Region of Xi'an; Table S5: Heat Supply Volume of Xi'an; Table S6: Building areas of various types in Xi'an; Table S7: China Heat Pump Market Sales Data; Table S8: Sales Growth Rate and Adoption Rate of China's Heat Pump Market; Table S9: Carbon emission prediction results of Xi'an from under the BAU scenario; Table S10: Carbon emission prediction results of Xi'an under the EMC scenario; Table S11: Carbon emission prediction results of Xi'an under the RFM scenario; Table S12: Carbon emissions reduction effect from building materials carbon sink; Table S13: Baseline scenario settings for heating and hot water supply; Table S14: Carbon emissions reduction effect from building heat pumps; Table S15: Carbon emissions reduction effect from rooftop photovoltaic systems; Table S16: Carbon emissions reduction effect from smart HVAC systems; Table S17: Carbon emissions reduction effect from building envelope improvement.

Author Contributions: Conceptualization, D.Z., and L.S.; methodology, D.Z., L.S., T.L., and L.G.; software, D.Z. and F.W.; validation, Y.Z., L.G., F.W., X.Q., and Y.L.; formal analysis, D.Z., Y.Z., L.G., and Y.L.; investigation, D.Z., L.G., X.Q., Y.L., and J.Z.; resources, D.Z., L.S.; data curation, D.Z., Y.Z., F.W., and J.Z.; writing—original draft preparation, D.Z., L.S., and Y.Z.; writing—review and editing, D.Z., L.S., and J.Z.; visualization, D.Z., F.W., and X.Q.; supervision, L.S., Y.W., and J.Z.; project administration, L.S.; funding acquisition, L.S., and L.G. All authors have read and agreed to the published version of the manuscript.

Funding: This research is supported by "The Youth Innovation Team of Shaanxi Universities", the Shanghai Pujiang Program (22PJC052), the Natural Science Foundation of Shaanxi (2023-JC-QN-0808; QCYRCXM-2022-127), and the National Science Foundation of Hunan Province (Grant No. 2023JJ40557).

Data Availability Statement: The original contributions presented in the study are included in the article; further inquiries can be directed to the corresponding author.

Conflicts of Interest: The authors declare no conflicts of interest.

References

1. Zhao, X.; Ma, X.; Chen, B.; Shang, Y.; Song, M. Challenges toward carbon neutrality in China: Strategies and countermeasures. *Resour. Conserv. Recycl.* **2022**, *176*, 105959. [CrossRef]
2. Ma, M.; Zhou, N.; Feng, W.; Yan, J. Challenges and opportunities in the global net-zero building sector. *Cell Rep. Sustain.* **2024**, *1*, 100154. [CrossRef]
3. Luo, W.; Liu, W.; Liu, W.; Xia, L.; Zheng, J.; Liu, Y. Analysis of influencing factors and carbon emission scenario prediction during building operation stage. *Energy* **2025**, *316*, 134401. [CrossRef]
4. Awan, A.; Kocoglu, M.; Subhan, M.; Utepkaliyeva, K.; bte Mohamed Yusoff, N.Y.; Hossain, M.E. Assessing energy efficiency in the built environment: A quantile regression analysis of CO2 emissions from buildings and manufacturing sector. *Energy Build.* **2025**, *338*, 115733. [CrossRef]
5. Shi, H.; Chai, J.; Lu, Q.; Zheng, J.; Wang, S. The impact of China's low-carbon transition on economy, society and energy in 2030 based on CO2 emissions drivers. *Energy* **2022**, *239*, 122336. [CrossRef]
6. Yang, J.; Deng, Z.; Guo, S.; Chen, Y. Development of bottom-up model to estimate dynamic carbon emission for city-scale buildings. *Appl. Energy* **2023**, *331*, 120410. [CrossRef]
7. Zhang, C.; Luo, H. Research on carbon emission peak prediction and path of China's public buildings: Scenario analysis based on LEAP model. *Energy Build.* **2023**, *289*, 113053. [CrossRef]
8. Tan, X.; Lai, H.; Gu, B.; Zeng, Y.; Li, H. Carbon emission and abatement potential outlook in China's building sector through 2050. *Energy Policy* **2018**, *118*, 429–439. [CrossRef]

9. Yuan, S.; Chen, X.; Du, Y.; Qu, S.; Hu, C.; Jin, L.; Xu, W.; Yan, G. Pathway of Carbon Emission Peak of China's Building Sector. *Res. Environ. Sci.* **2022**, *35*, 394–404. [CrossRef]
10. Xu, W.; Sun, D.; Dong, J.; Huang, S.; Yang, S. Technical pathways of dual carbon control in China's buildings sector. *Carbon Neutrality* **2022**, *1*, 18. [CrossRef]
11. Li, S.; Kong, W.; Wang, Y.; Yuan, L. Medium and long-term energy demand forecasts by sectors in China under the goal of "carbon peaking & carbon neutrality": Based on the LEAP-China model. *Energy* **2024**, *310*, 133017. [CrossRef]
12. Yuan, Y.; Wang, F.; Sun, L.; Liu, W.; Du, C.; Wang, C.; Yao, Y. Promote the decarbonization pathways of eco-industrial parks by considering their CO2 emissions structures and characters. *J. Clean. Prod.* **2024**, *450*, 141989. [CrossRef]
13. Núñez, E.; Steyerberg, E.W.; Núñez, J. Regression Modeling Strategies. *Rev. Española De Cardiol.* **2011**, *64*, 501–507. [CrossRef]
14. Churkina, G.; Organschi, A.; Reyer, C.P.O.; Ruff, A.; Vinke, K.; Liu, Z.; Reck, B.K.; Graedel, T.E.; Schellnhuber, H.J. Buildings as a global carbon sink. *Nat. Sustain.* **2020**, *3*, 269–276. [CrossRef]
15. Zhao, R.; Huang, X.; Xue, J.; Guan, X. A practical simulation of carbon sink calculation for urban buildings: A case study of Zhengzhou in China. *Sustain. Cities Soc.* **2023**, *99*, 104980. [CrossRef]
16. Gogtay, N.J.; Thatte, U.M. Principles of correlation analysis. *J. Assoc. Physicians India* **2017**, *65*, 78–81.
17. Cleophas, T.J.; Zwinderman, A.H. Bayesian Pearson Correlation Analysis. In *Modern Bayesian Statistics in Clinical Research*; Cleophas, T.J., Zwinderman, A.H., Eds.; Springer International Publishing: Cham, Switzerland, 2018; pp. 111–118. [CrossRef]
18. Zhang, Y.; Zheng, W.; Fang, H.; Xia, J. Clean heating in Northern China: Regional investigations and roadmap studies for urban area towards 2050. *J. Clean. Prod.* **2022**, *334*, 130233. [CrossRef]
19. Xu, Z.; Li, H.; Xu, W.; Shao, S.; Wang, Z.; Gou, X.; Zhao, M.; Li, J. Investigation on the efficiency degradation characterization of low ambient temperature air source heat pump under partial load operation. *Int. J. Refrig.* **2022**, *133*, 99–110. [CrossRef]
20. Zhong, T.; Zhang, Z.; Chen, M.; Zhang, K.; Zhou, Z.; Zhu, R.; Wang, Y.; Lü, G.; Yan, J. A city-scale estimation of rooftop solar photovoltaic potential based on deep learning. *Appl. Energy* **2021**, *298*, 117132. [CrossRef]
21. Shafique, M.; Luo, X.; Zuo, J. Photovoltaic-green roofs: A review of benefits, limitations, and trends. *Sol. Energy* **2020**, *202*, 485–497. [CrossRef]
22. Fakhraian, E.; Forment, M.A.; Dalmau, F.V.; Nameni, A.; Guerrero, M.J.C. Determination of the urban rooftop photovoltaic potential: A state of the art. *Energy Rep.* **2021**, *7*, 176–185. [CrossRef]
23. Peng, J.; Lu, L. Investigation on the development potential of rooftop PV system in Hong Kong and its environmental benefits. *Renew. Sustain. Energy Rev.* **2013**, *27*, 149–162. [CrossRef]
24. Yan, J.; Yang, Y.; Elia Campana, P.; He, J. City-level analysis of subsidy-free solar photovoltaic electricity price, profits and grid parity in China. *Nat. Energy* **2019**, *4*, 709–717. [CrossRef]
25. Wang, Y.; He, J.; Chen, W. Distributed solar photovoltaic development potential and a roadmap at the city level in China. *Renew. Sustain. Energy Rev.* **2021**, *141*, 110772. [CrossRef]
26. Wu, W.; Skye, H.M.; Domanski, P.A. Selecting HVAC systems to achieve comfortable and cost-effective residential net-zero energy buildings. *Appl. Energy* **2018**, *212*, 577–591. [CrossRef]
27. Homod, R.Z. Analysis and optimization of HVAC control systems based on energy and performance considerations for smart buildings. *Renew. Energy* **2018**, *126*, 49–64. [CrossRef]
28. Idrissi Kaitouni, S.; Sangkyu, P.; Mghazli, M.O.; El Mansouri, F.; Jamil, A.; Ahachad, M.; Brigui, J. Design parameters influencing the energy performance and indoor comfort of net zero energy building "NZEB" designed for semi-arid urban areas: Digital workflow methodology, sensitivity analysis and comparative assessment. *Sol. Energy* **2024**, *268*, 112264. [CrossRef]
29. Huang, J.; Wang, S.; Teng, F.; Feng, W. Thermal performance optimization of envelope in the energy-saving renovation of existing residential buildings. *Energy Build.* **2021**, *247*, 111103. [CrossRef]
30. Zhao, B.; Lu, K.; Zhang, W.; Jin, C.; Xuan, Q.; Pei, G. Thermo-responsive hydrogel-based building envelopes for building energy-saving. *Sol. Energy* **2025**, *288*, 113306. [CrossRef]
31. Qin, X.; Liu, H.; Zhang, X.; Jiang, N.; Yang, L.; Jin, X. Thermal analysis of the window-wall interface for renovation of historical buildings. *Energy Build.* **2024**, *310*, 114108. [CrossRef]
32. Bai, L.; Wang, S. Definition of new thermal climate zones for building energy efficiency response to the climate change during the past decades in China. *Energy* **2019**, *170*, 709–719. [CrossRef]
33. Liu, C.; Ma, W.; Hao, J.; Luo, D.; Zuo, J.; Zhang, C. Energy Retrofitting Assessment of Public Building Envelopes in China's Hot Summer and Cold Winter Climate Region. *Buildings* **2022**, *12*, 1866. [CrossRef]
34. Zheng, Z.; Zhou, J.; Yang, Y.; Xu, F.; Liu, H. Economic optimization of exterior wall insulation in Chinese office buildings by coupling artificial neural network and genetic algorithm. *Therm. Sci. Eng. Prog.* **2024**, *50*, 102582. [CrossRef]
35. Ren, W.R.; Wang, C.H.; Han, C.B.; Han, D.; Zheng, J.Y.; Cui, Y.N.; Song, X.M.; Jiang, Q.; Yan, H. Cement-based external wall insulation material with thermal performance improvement by partial substitution of calcium silicate. *Energy Build.* **2024**, *317*, 114415. [CrossRef]

36. Mata, É.; Korpal, A.K.; Cheng, S.H.; Jiménez Navarro, J.P.; Filippidou, F.; Reyna, J.; Wang, R. A map of roadmaps for zero and low energy and carbon buildings worldwide. *Environ. Res. Lett.* **2020**, *15*, 113003. [CrossRef]
37. Lippiatt, N.; Ling, T.-C.; Pan, S.-Y. Towards carbon-neutral construction materials: Carbonation of cement-based materials and the future perspective. *J. Build. Eng.* **2020**, *28*, 101062. [CrossRef]
38. Lu, H.; You, K.; Feng, W.; Zhou, N.; Fridley, D.; Price, L.; de la Rue du Can, S. Reducing China's building material embodied emissions: Opportunities and challenges to achieve carbon neutrality in building materials. *iScience* **2024**, *27*, 109028. [CrossRef]

buildings

MDPI

Article

The Coupling Coordination Degree and Spatio-Temporal Divergence Between Land Urbanization and Energy Consumption Carbon Emissions of China's Yangtze River Delta Urban Agglomeration

Zhengru Li, Yang Yu *, Bo Liu, Xiaoyu Zhang, Tianyin Li, Nuo Shi and Yichen Ren

College of Architecture, Xi'an University of Architecture and Technology, Xi'an 710054, China
* Correspondence: yuyang@xauat.edu.cn; Tel.: +86-15802911789

Abstract: The strategic coordinated development of land urbanization and carbon emission systems in urban agglomerations is crucial for achieving dual carbon goals and sustainable development. While existing studies emphasize population and economic urbanization, the spatiotemporal coupling mechanisms between land urbanization (encompassing size, input, and output dimensions) and carbon emissions remain underexplored. This study collects data on land urbanization and carbon emissions from 27 cities in China's Yangtze River Delta urban agglomeration between 2010 and 2019. By establishing evaluation systems for land urbanization and energy consumption carbon emission subsystems, by and employing coupling coordination degree models with spatial autocorrelation analysis methods, this paper analyzes the spatiotemporal dynamic evolution characteristics of the coupled coordination relationship between land urbanization and energy consumption carbon emissions in the Yangtze River Delta urban agglomeration. The results indicate the following: (1) From 2010 to 2019, the comprehensive level of the land urbanization subsystem in the Yangtze River Delta urban agglomeration continued to rise, with higher comprehensive indices in the southern and northern peripheral regions and lower values in central urban areas. The carbon emission subsystem showed sustained stable decline, with a gradual reduction in the number of cities maintaining low carbon emission levels. (2) Temporally, the overall coupling coordination degree of the urban agglomeration system demonstrated an upward trend, progressing from severe imbalance to the primary coordination stage. (3) Spatially, significant regional differences in coupling coordination degree were observed, showing higher values in the southeastern areas compared to the northwestern regions. (4) Most areas exhibited no significant clustering characteristics in the coupling coordination degree between land urbanization and energy consumption carbon emissions, while the local spatial clustering patterns demonstrated temporal variations. These findings systematically reveal the transition mechanisms of land–carbon coordination in urban agglomerations, providing empirical evidence to resolve the theoretical debate on urbanization's dual role in emission promotion and reduction.

Keywords: land urbanization; energy consumption carbon emissions; coupling coordination degree; spatial autocorrelation; spatiotemporal dynamic evolution

1. Introduction

The Intergovernmental Panel on Climate Change (IPCC) reports that over 70% of global carbon emissions originate from cities [1]. Notably, urban areas are critical target regions for reducing global carbon emissions. In China, national development strategies such as new-type urbanization have further promoted the spatial concentration of regional economies. Balancing urban economic growth with emission reduction targets poses a severe challenge for China's future. China's '14th Five-Year Plan for Modern Energy Systems' explicitly aims to reduce carbon emissions per unit of GDP by 18% and energy consumption per unit of GDP by 13.5% cumulatively by 2025. Meanwhile, for both developed and emerging economies, urbanization has entered the era of urban agglomerations, which have become focal areas for a range of environmental issues. As economic scales expand, carbon emission pressures will continue to rise [2]. In 2023, the 28th Conference of the Parties (COP28) to the United Nations Framework Convention on Climate Change (UNFCCC) was held in Dubai, UAE. During the conference, the China Pavilion hosted a themed side event titled "Narrating China's Climate Action Stories and Global Climate Governance for Low-Carbon Cities and Urban Agglomerations", which jointly explored pathways for carbon peaking and carbon neutrality in cities and urban agglomerations. The coordination between urbanization and carbon emission systems in urban agglomerations is essential for advancing low-carbon sustainable urbanization and planning. Consequently, studying the development and emission reduction patterns of representative urban agglomerations holds significant reference value for other regions globally [3–6].

In recent years, the relationship between urbanization and carbon emissions has attracted considerable academic attention. It is widely recognized that urbanization, as a critical driver of energy consumption and economic growth, is closely associated with carbon emissions [7–14]. However, most existing studies focus predominantly on population urbanization and economic urbanization, while land urbanization, which constitutes the physical and spatial foundation of urban expansion, has received comparatively limited attention. For instance, Xu et al. [15] incorporated economic, social, and ecological benefits into their evaluation of land urbanization quality and found a suppressive effect on carbon emissions. Tang et al. [16] further identified land-use intensity, structure, and efficiency as key positive drivers of emissions. Nevertheless, the impact of land urbanization on carbon emissions remains debatable, with findings varying across different development stages and regional contexts [17–19]. Moreover, most existing analyses are conducted at national or provincial levels, lacking detailed examination of the structural characteristics of land urbanization and their carbon implications at the urban agglomeration scale [20–22].

Urban agglomerations, as major hubs of economic activity and energy demand, have become critical focal points for carbon emission research [23–33]. Liu et al. [34] found that spatial structure within agglomerations significantly affects emissions: spatial concentration and transportation accessibility tend to increase emission efficiency, whereas compact urban form helps mitigate emissions. Dong et al. [35] revealed the short-term inhibitory effects of land-use efficiency on carbon intensity in the Yangtze River Delta through stochastic frontier and entropy-based models. Zhou et al. [36] estimated land-use carbon emissions in the Beijing-Tianjin-Hebei region, highlighting built-up land expansion as a key contributor to increasing emissions. Although these studies have deepened understanding of emissions in urban clusters, they seldom construct comprehensive index systems for evaluating land urbanization–carbon interactions, nor do they systematically analyze spatial coordination mechanisms [22,37,38].

With the advancement of spatial econometric methodologies, recent studies have increasingly emphasized spatial heterogeneity and spillover effects in carbon emissions research [39–41]. Xu et al. [39] employed the Environmental Kuznets Curve (EKC) framework in the Pearl River Delta to reveal that economic urbanization exerts the strongest impact on emissions, followed by land and population urbanization. Chen et al. [40] used spatial autocorrelation and panel models to analyze multidimensional urbanization effects, identifying sustained emission increases in areas with rapid land expansion. Zhu et al. [41] showed that land fragmentation exacerbates emissions, while improved urban land connectivity supports mitigation. However, these approaches primarily rely on regression-based correlations, which fall short in capturing the interactive and dynamic coordination between systems. Particularly lacking are models that can evaluate mutual interactions and co-evolutionary trends across multiple urban systems, especially within spatially integrated agglomerations.

In summary, although considerable progress has been made in investigating the urbanization–carbon emission nexus—especially regarding multidimensional urbanization and urban agglomeration scales [7,10,15,16,33,35,40]—several critical research gaps persist. First, current studies rarely conduct systematic analyses of the internal structure of land urbanization, including its scale, investment, and output dimensions [18,20,22]. Second, the intercity heterogeneity within urban agglomerations is often overlooked, limiting the understanding of spatial coordination and spillover patterns [32,34,35,37]. Third, most analyses are based on static or linear models, which are inadequate for capturing the spatiotemporal evolution and coordination mechanisms between land use and emissions [39–41]. Therefore, there is a pressing need for integrated quantitative frameworks that evaluate both interaction strength and coordination trends, supplemented by spatial diagnostic tools to assess regional synergy.

To address these gaps, this study focuses on the Yangtze River Delta, China's most representative urban agglomeration. Using panel data from 27 cities from 2010 to 2019, a comprehensive evaluation index system is constructed, encompassing land urbanization indicators (scale, input, output) and energy-related carbon emissions. A Coupling Coordination Degree (CCD) model is employed to assess the interactive dynamics between these two systems, while global and local spatial autocorrelation (e.g., Moran's I) are used to analyze the spatial clustering and evolution patterns [41,42]. Compared to traditional methods such as LMDI decomposition, IPAT/Kaya identity models, and regression analysis [22,37], the CCD approach offers significant advantages: (1) it quantifies the degree and quality of system interaction and coordination over time, which is critical for multi-system studies [20,21,43,44]; and (2) it is well-suited for integration with GIS and spatial econometric tools, enabling dynamic spatial pattern analysis and comparison [40,42,45,46]. The key innovations of this study are threefold: (1) it provides the first systematic quantification of the coordinated dynamics between land urbanization and carbon emissions; (2) it addresses spatial imbalance within urban agglomerations, revealing spatiotemporal coordination disparities; and (3) it integrates methodological innovation and spatial-scale precision to offer theoretical and practical guidance for achieving coordinated low-carbon urbanization.

2. Study Area and Data Sources

2.1. Study Area

The Yangtze River Delta urban agglomeration is located in the lower reaches of the Yangtze River, spanning 114°54′–123°10′ E and 27°02′–35°08′ N, adjacent to the Yellow Sea and East China Sea. It forms an alluvial plain before the Yangtze River flows into the sea. The urban agglomeration covers Shanghai, Jiangsu, Zhejiang, and Anhui provinces, encompassing 27 cities with a total area of 225,000 km^2. By 2021, the average urbanization rate in the Yangtze River Delta urban agglomeration exceeded 68%, and its GDP reached CNY 27.3 billion. As a critical intersection of China's "Belt and Road" initiative and the Yangtze River Economic Belt, it holds strategic significance in national modernization. The region heavily relies on fossil fuels such as coal, oil, and natural gas, with total primary energy demand reaching approximately 840 million tons of standard coal in 2020, facing substantial challenges in low-carbon sustainable development and energy transition.

2.2. Data Sources

Following the Outline of the Yangtze River Delta Regional Integration Development Plan issued by the Central Committee of the Communist Party of China and the State Council in 2019, this study focuses on 27 core cities in the Yangtze River Delta from 2010 to 2019. These include Shanghai, nine cities in Jiangsu (Nanjing, Zhenjiang, Yangzhou, Changzhou, Suzhou, Wuxi, Nantong, Taizhou, Yancheng), nine cities in Zhejiang (Hangzhou, Jiaxing, Huzhou, Shaoxing, Ningbo, Wenzhou, Zhoushan, Jinhua, Taizhou), and eight cities in Anhui (Hefei, Wuhu, Chuzhou, Ma'anshan, Tongling, Chizhou, Anqing, Xuancheng). Data were sourced from the China National Bureau of Statistics (2010–2019), *China Urban Construction Statistics Yearbook* (2010–2019), and Carbon Emission Accounts and Datasets (CEADs) (2010–2019) (Table 1). Missing values were predicted and supplemented using linear regression [38,47]. Data for all 27 cities from 2010 to 2019 were systematically compiled and standardized for analysis.

Table 1. Data sources and corresponding references.

Database	Repository Name	Access Link	Key Citations
1	China National Bureau of Statistics	https://data.stats.gov.cn/index.htm (accessed on 25 May 2025).	-
2	*China Urban Construction Statistics Yearbook*	https://www.mohurd.gov.cn/gongkai/fdzdgknr/sjfb/tjxx/jstjnj/index.html (accessed on 25 May 2025).	-
3	Carbon Emission Accounts and Datasets (CEADs)	https://www.ceads.net/user/index.php?id=1281&lang=cn (accessed on 25 May 2025).	[48–51]

3. Methods

3.1. Evaluation Indicator System Construction

First, this study defines the concept of land urbanization [52]. Second, principles of scientific rigor and data availability are applied to integrate land urbanization and energy consumption carbon emissions. Third, existing evaluation index systems for land urbanization at the urban agglomeration level from the domestic and international literature are referenced. Finally, a coupled coordination degree index system for land urbanization and energy consumption carbon emissions in the Yangtze River Delta urban agglomeration is constructed (Table 2). The system includes aggregate indicators, reflecting regional-scale and relative indicators and representing average levels across regions to mitigate inaccuracies caused by differences in regional scope and size.

Table 2. Evaluation index system of CCD between the urbanization of urban land and the carbon emissions of energy consumption in the Yangtze River Delta city cluster.

System	Subsystem	Indicators	The Measuring Unit	Index Effect
land urbanization	Land size level [15,16,43,53–56]	Urban built-up area	km^2	Positive
		Per capita public green space	Persons/m^2	Positive
	Land input level [15,42,57]	Government expenditure per land	Ten thousand yuan/km^2	Positive
		Fixed asset investment in municipal public facilities construction per land	Ten thousand yuan/km^2	Positive
	Land output level [15,22]	GDP per unit land area	Ten thousand yuan/km^2	Positive
		Land average output value of secondary and tertiary industries	Ten thousand yuan/km^2	Positive
carbon emission	Carbon emission level [49–51,55]	CO_2 emissions from the production of 17 fossil fuels and cement	Ten thousand tons standard coal	Negative

3.2. Framework of Coupling Coordination Analysis

Based on panel data from 27 cities in the Yangtze River Delta (2010–2019), this study first employs the entropy method to calculate the comprehensive indices of land urbanization and carbon emissions. Subsequently, a CCD model is introduced to compute system and subsystem CCD. Finally, global and local spatial autocorrelation methods are applied to analyze the spatiotemporal dynamic evolution characteristics of the coupled coordination relationship between land urbanization and energy consumption carbon emissions in the Yangtze River Delta urban agglomeration (Figure 1).

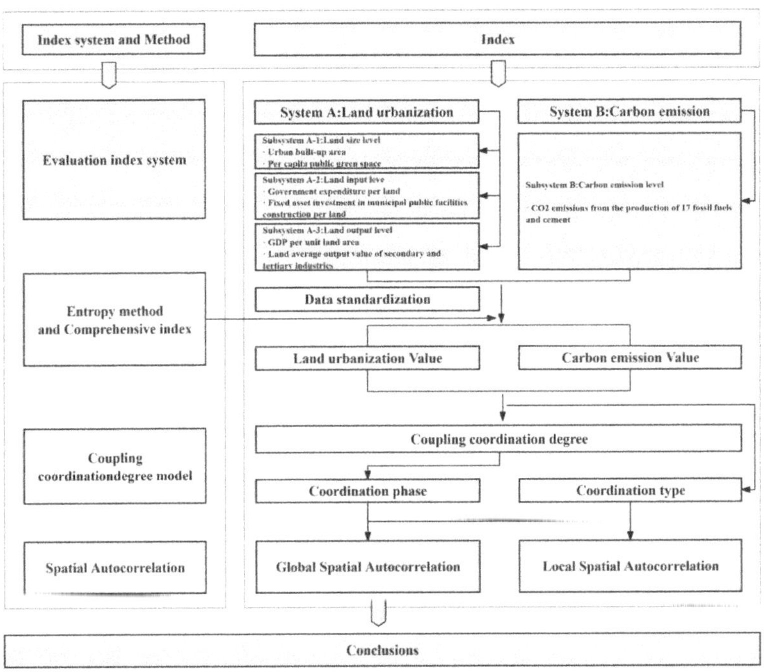

Figure 1. Overall research framework.

3.2.1. Entropy Method

Entropy is a measure of the uncertainty of the state or of its average information content [58]. Shannon stated that the entropy can be interpreted as the average rate at which information is produced by a stochastic source of data. When the data source produces a low-entropy value, the event carries more "information". The entropy method is an objective and comprehensive weighting method which is based on the dispersion degree of the evaluation index data to measure the index weight [59]. As an objective weight measurement method, the entropy weight method is widely used in the assessment of urbanization and sustainable development level [44,60]. This study utilizes information entropy to calculate indicator weights, providing a basis for analyzing the comprehensive indices of land urbanization and carbon emissions in the Yangtze River Delta urban agglomeration.

For the entropy method, negative-effect indicators require forward processing. The extremum method is used to standardize raw indicator values. The standardization formulas for positive-benefit and negative-benefit indicators are as follows [16,38,60]:

$$x'_{ij} = \begin{Bmatrix} (x_{ij} - x_{jmin})/(x_{jmax} - x_{jmin}) \\ (x_{jmax} - x_{ij})/(x_{jmax} - x_{jmin}) \end{Bmatrix}, i = 1, 2, \cdots, m; j = 1, 2, \cdots, n \tag{1}$$

where x'_{ij} is the standardized information value, x_{ij} is the original value of the jth indicator in year i; m is the number of years; and n is the number of evaluation indicators.

The information entropy and utility values of the metrics are further calculated with the following formula:

$$P_{ij} = x_{ij} / \sum_{i=1}^{m} x_{ij} \tag{2}$$

$$e_j = -(1/\ln(m)) \times \sum_{i=1}^{m} P_{ij} \ln(P_{ij}) \tag{3}$$

$$g_i = 1 - e_j \tag{4}$$

The final calculation of the weights of the indicators is based on the following formula:

$$\omega_j = g_j / \sum_{j=1}^{n} g_j \tag{5}$$

3.2.2. Comprehensive Index

The comprehensive index is an evaluation approach that synthesizes and reflects multiple indicators [61]. This study adopts the standard deviation normalization method to calculate the comprehensive indices of land urbanization and carbon emissions in the Yangtze River Delta urban agglomeration [20]. This method enables the measurement of comprehensive indices for the land urbanization system, energy consumption carbon emission system, and their subsystems, laying the foundation for subsequent CCD analysis. The calculation formula is as follows [20,62]:

$$Y_k = \sum_{j=1}^{n} \omega_j \times x'_{ij}, k = 1, 2, \cdots, 27 \tag{6}$$

where Y_k represents the comprehensive index, ω_j denotes indicator weights, and x'_{ij} is the standardized value of the j-th indicator for the i-th sample.

$$Y_{k-t} = \omega \times Y_k \tag{7}$$

$$\omega = t / \sum_{t=1}^{m} t, t = 1, 2, \ldots, m \tag{8}$$

Based on this, a comprehensive evaluation and analysis are conducted. A higher comprehensive index indicates a higher level of urbanization development, while a lower index suggests the opposite. By combining the variance of ranking results, the overall volatility of the sample is assessed. A smaller ranking variance implies lower volatility in the urban agglomeration's development, whereas a larger variance reflects higher volatility.

3.2.3. Coupling Coordination Degree Model

The concept of "coupling" originated in physics, referring to the interaction between systems. It has since been applied to studies on economic development, industrial structure, social environments, and technological innovation [63]. Coupling degree measures the intensity of interaction between systems, such as land urbanization and energy consumption carbon emissions, to assess their temporal development order [21,64]. Coordination degree, on the other hand, reflects the level of mutual alignment between systems within the same period, evaluating their harmonious consistency.

This study defines the CCD as the extent to which land urbanization and energy consumption carbon emissions interact and influence each other through their respective elements. The model comprises three components: coordination degree (T), coupling degree (C), and CCD, with classification criteria and thresholds.

(1) Coordination Degree

The coordination degree represents the comprehensive evaluation index of the development levels of land urbanization and energy consumption carbon emissions, reflecting their overall synergistic benefits. The formula is as follows [47,62]:

$$T = a\mu_1 + b\mu_2 \tag{9}$$

In the formula, a and b are the weights to be determined, and this study considers that the contribution of land urbanization and energy consumption carbon emissions were the same, so a and b take the same weight value of 0.5 [16,60]; μ_1 and μ_2 represent land urbanization and energy consumption carbon emissions, respectively.

(2) Coupling Degree

Drawing from the physics concept of "capacity coupling" and the coupling coefficient model, the coupling degree quantifies the interdependence between systems. For land urbanization and energy consumption carbon emissions, the formula is as follows [60]:

$$C = \left[(U_1 \times U_2 \times U_3 \times \cdots U_n)/(U_i + U_j)\right]^{1/m} \tag{10}$$

$$C = 2\left[(\mu_1 \times \mu_2)/(\mu_1 + \mu_2)^2\right]^{1/2} \tag{11}$$

where μ (n = 1, 2, … m) in each subsystem assessment value, m was the number of subsystems, and because this study only uses 2 systems of land urbanization and energy consumption carbon emissions, m = 2. C was the coupling degree, and C ∈ [0, 1] [20], if the larger the value of C, the more coupled the two developments are, C = 1, to reach the best coupling state.

(3) Coupling Coordination Degree

Relying solely on T or C may introduce errors, obscure dynamic trends, and hinder cross-sectional comparisons. Thus, the CCD integrates both metrics to measure the coordinated development status [21,60,62]:

$$CCD = \sqrt{C \times T} \tag{12}$$

where CCD ∈ [0, 1] [20].

(4) Classification Standards

Based on existing studies, CCD values are classified into distinct coordination levels (Table 3) [53].

Table 3. Stage division of CCD.

The Serial Number	Degree of Coupling Coordination	Coordinated Coupling Degree Stage
1	0.0 < CCD ≤ 0.1	Stage of extreme disorder
2	0.1 < CCD ≤ 0.2	Stage of severe disorder
3	0.2 < CCD ≤ 0.3	Stage of moderate disorder
4	0.3 < CCD ≤ 0.4	Stage of mild disorder
5	0.4 < CCD ≤ 0.5	Stage of near disorders
6	0.5 < CCD ≤ 0.6	Stage of grudging coordination
7	0.6 < CCD ≤ 0.7	Stage of primary coordination
8	0.7 < CCD ≤ 0.8	Stage of intermediate coordination
9	0.8 < CCD ≤ 0.9	Stage of good coordination
10	0.9 < CCD ≤ 1.0	Stage of quality coordination

3.2.4. Spatial Autocorrelation

Spatially, the CCD between land urbanization and energy consumption carbon emissions in the Yangtze River Delta exhibits significant spatial correlations. To analyze these patterns, this study employs Exploratory Spatial Data Analysis (ESDA), a suite of spatial statistical methods to assess spatial dependencies [41,65]. Global Spatial Autocorrelation: Measured using Moran's I index, this evaluates the overall spatial clustering of the observed indicator (e.g., CCD) across the study area. Local Spatial Autocorrelation: Assessed via Local Moran's I index, this identifies spatial clusters or outliers and quantifies the association between each spatial unit and its neighbors [40,42]. Its calculation formula is as follows [23]:

$$Moran's\ I = \frac{n\sum_i \sum_{i \neq j} wij(yi - \overline{y})(yj - \overline{y})}{(\sum_i \sum_{i \neq j} wij)\sum_i (yi - y)^2} \tag{13}$$

where n is the number of spatial units, wij is the value in the spatial weights matrix w corresponding to the geographic districts, and y is the value of CCD in this paper. The value of $Moran's\ I$ ranges from -1 to 1. A positive value of $Moran's\ I$ means the positive spatial correlation among the spatial units; a negative value of $Moran's\ I$ indicates negative spatial correlation, and 0 implies no spatial correlation. Statistical tests of spatial autocorrelation can be judged by the z-value of $Moran's\ I$. High z-values indicate highly significant spatial correlation.

Using ArcGIS 10.8 (Esri, Redlands, CA, USA) and GeoDa software 1.20 (GeoDa Center, Tempe, AZ, USA), this study analyzes the spatial autocorrelation of the CCD between land urbanization and energy consumption carbon emissions in the Yangtze River Delta urban agglomeration.

4. Results

4.1. Analysis of the Land Urbanization and Carbon Emissions Comprehensive Index

To investigate the comprehensive development level of the Yangtze River Delta urban agglomeration from 2010 to 2019, this study employs the inter-group addition and intra-group multiplication calculation method based on the established evaluation system to analyze the multi-indicator comprehensive indices at the urban agglomeration level, as shown in Figure 2. The comprehensive index of the Yangtze River Delta urban agglomeration system increased from 86.442 in 2010 to 144.164 in 2019, with an average annual growth rate of 5.772%. The comprehensive index of the land urbanization subsystem rose from 67.471 in 2010 to 128.804 in 2019, with an average annual growth rate of 4.762%, indicating an overall positive upward trend in the land urbanization and carbon emission systems. Meanwhile, the comprehensive index of the carbon emission subsystem decreased from 18.971 in 2010 to 15.36 in 2019, with an average annual decline rate of 1.903%. These results demonstrate that the overall system exhibited a positive upward trend, with the comprehensive indices of the system and the land urbanization subsystem showing highly similar growth trajectories (annual growth rate difference: 1.01%). This suggests that the system's comprehensive index was predominantly influenced by the land urbanization subsystem, while the carbon emission subsystem had a minimal impact on the system's overall index.

Following the analysis of the comprehensive indices at the urban agglomeration level, this study further examines the comprehensive indices of the 27 cities within the Yangtze River Delta urban agglomeration from 2010 to 2019 (Figure 3). Based on their comprehensive index growth rates, the cities are categorized into four development levels from low to high. Cities with growth rates between 0% and 0.3% include Wuxi, Suzhou, Zhenjiang, Zhoushan, Tongling, Ma'anshan, and Chizhou, totaling seven cities. Cities with growth rates ranging from 0.3% to 0.6% encompass Shanghai, Nanjing, Changzhou, Nantong, Yangzhou, Yancheng, Taizhou, Ningbo, Wuhu, Anqing, and Xuancheng, comprising 11 cities. Cities with growth rates of 0.6% to 1.0% consist of Wenzhou, Huzhou, Shaoxing, Jinhua, and Hefei, totaling five cities. Cities with growth rates exceeding 1.0% include Hangzhou, Jiaxing, Taizhou, and Chuzhou, amounting to four cities. Overall, significant progress in comprehensive indices was observed across all 27 cities during the study period.

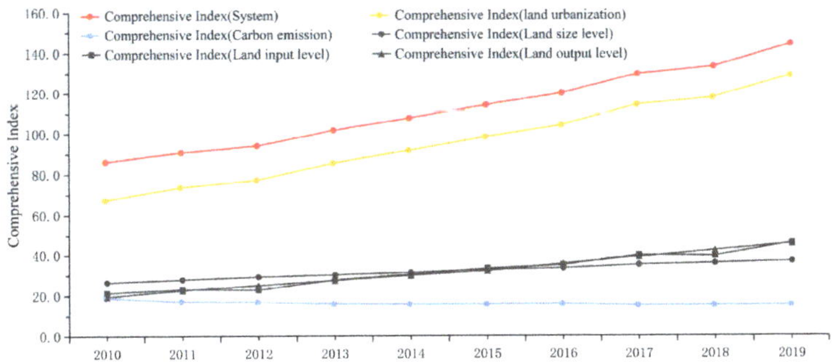

Figure 2. Comprehensive index analysis of land urbanization and carbon emissions in the Yangtze River Delta urban agglomeration, China, 2010–2019.

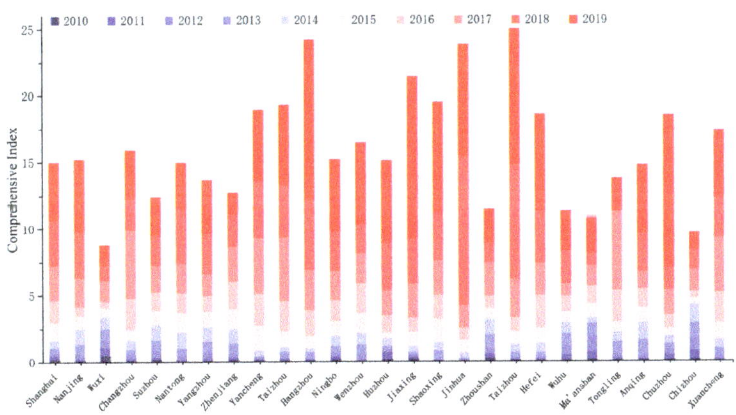

Figure 3. Comprehensive index analysis of land urbanization and carbon emission system of energy consumption in 27 cities in the Yangtze River Delta, China, 2010–2019.

Following the analysis of comprehensive indices at both the urban agglomeration and individual city levels, this study further examines the comprehensive indices of the land urbanization subsystem and carbon emission subsystem across the 27 cities in the Yangtze River Delta urban agglomeration from 2010 to 2019 (Figure 4a,b). From 2010 to 2019, the comprehensive index of land urbanization consistently exceeded that of carbon emissions. The smallest gap between the two indices occurred in 2010 (48.5), while the largest gap reached 113.444 in 2019, with an annual average difference of 79.7659. These results highlight a persistent imbalance in the development between the land urbanization and carbon emission subsystems.

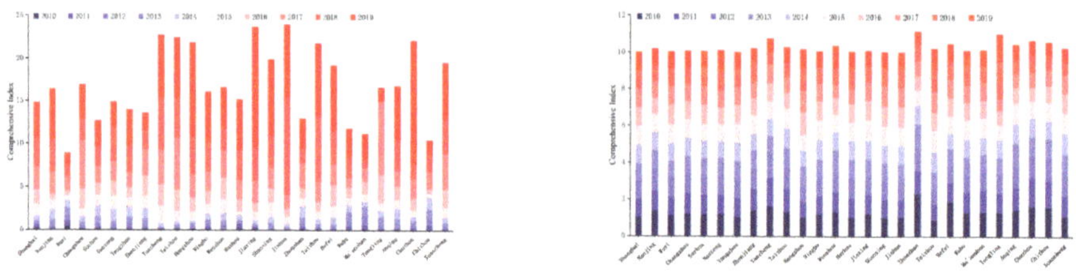

(a) (b)

Figure 4. (**a**) Comprehensive index analysis of land urbanization subsystem of 27 cities in the Yangtze River Delta city cluster, China, 2010–2019. (**b**) Comprehensive index analysis of carbon emission subsystem of 27 cities in the Yangtze River Delta city cluster, China, 2010–2019.

4.2. Time Variation Characteristics of the Coupled Coordination Between the Land Urbanization and Carbon Emissions

Based on the calculation steps of the CCD model, including data processing and weight determination, the CCD between land urbanization and carbon emissions in the Yangtze River Delta urban agglomeration from 2010 to 2019 was derived (Figure 5). The system's CCD increased from 0.195 (severe imbalance) in 2010 to 0.541 (marginal coordination) in 2019, demonstrating an overall upward trend. Subsystem analysis revealed divergent trajectories: (1) The CCD of the land urbanization subsystem rose steadily from 0.1 (extreme imbalance) in 2010 to 0.995 (high-quality coordination) in 2019. (2) The CCD of the carbon emission subsystem declined from 0.902 (high-quality coordination) in 2010 to 0.296 (moderate imbalance) in 2019. (3) The CCD trends of the two subsystems exhibited opposite directions. The largest gap between subsystem CCD values occurred in 2010 (0.098), coinciding with the lowest system CCD (0.195). The smallest gap (0.057) was observed in 2013, when the system CCD reached a relatively high level (0.605). For secondary indicators of land urbanization (land scale, land input, and land output), their CCD values followed similar upward trajectories. Overall, these findings indicate a continuous improvement in the coordination between land urbanization and carbon emissions over the decade, with low-carbon urbanization gradually progressing in a positive direction.

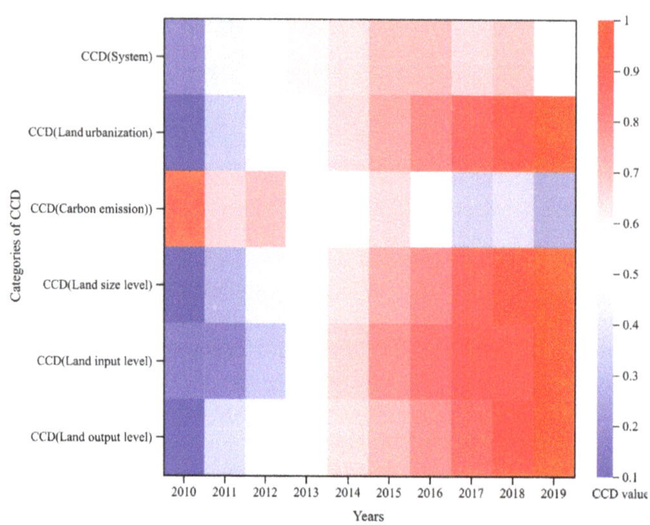

Figure 5. Analysis of the system-level CCD between land urbanization and carbon emissions from energy consumption in 27 cities of the Yangtze River Delta urban agglomeration in China, 2010–2019.

4.3. Spatial Variation in CCD Between Land Urbanization and Carbon Emissions

Based on ArcGIS 10.4 software, the spatial distribution of CCD across the 27 cities in the Yangtze River Delta urban agglomeration from 2010 to 2019 is illustrated in Figure 6. In 2010, all cities were categorized into extreme imbalance or severe imbalance stages. By 2019, most cities transitioned into various coordination stages. Only Zhenjiang, Chuzhou, and Wuxi (three cities) remained in the marginal coordination stage, while the rest advanced to primary coordination or higher. The number of cities in high coordination stages gradually increased: Shanghai, Huzhou, Taizhou, and Wenzhou (four cities) reached the high-quality coordination stage, and Nanjing, Hefei, Anqing (ten cities) achieved the good coordination stage. Cities in high-quality coordination were predominantly concentrated in the eastern and southern coastal regions, whereas those in good coordination were mainly located in the western regions. Notably, Shanghai transitioned from marginal coordination in 2017 to high-quality coordination in 2018, while cities such as Wuhu, Zhenjiang, and Shaoxing experienced minor regressions. The analysis reveals significant growth trends in CCD across all cities, although notable disparities persist among them.

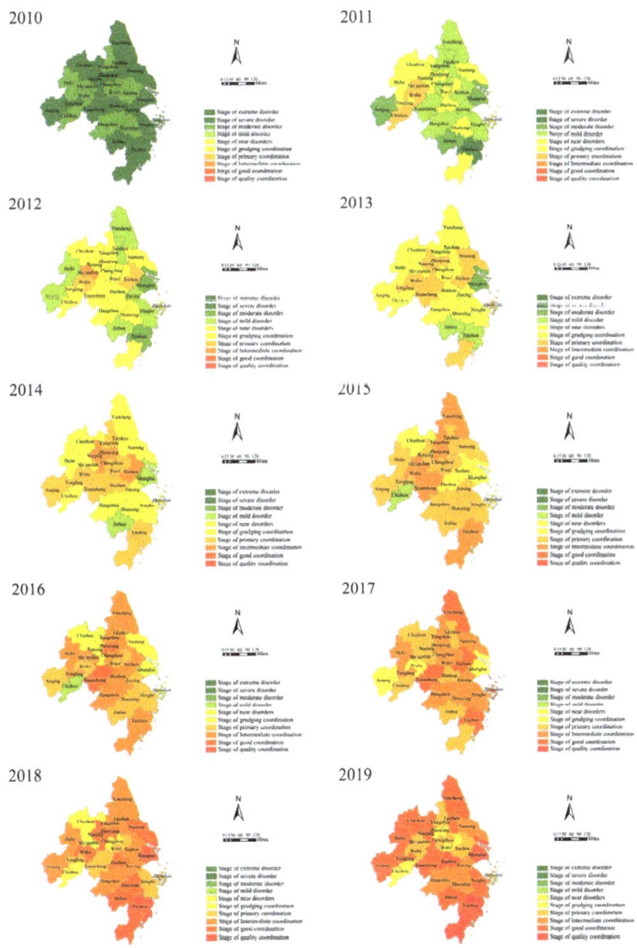

Figure 6. Spatial distribution of the CCD between land urbanization and energy consumption and carbon emission in the Yangtze River Delta city cluster, 2010–2019.

4.4. Spatial Autocorrelation Analysis of Land Urbanization and Carbon Emissions

4.4.1. Global Spatial Autocorrelation

Using panel data from the 27 cities within the Yangtze River Delta urban agglomeration (2010–2019), this study calculates the Global Moran's I index for the CCD between land urbanization and energy consumption carbon emissions using GeoDa software 1.20 (GeoDa Center, Tempe, AZ, USA) (Figure 7). The Moran's I values alternated between negative and positive during the study period, indicating unstable spatial correlations in CCD levels. This fluctuation reflects heterogeneous coordination levels among neighboring cities, with unstable interdependencies and spatially clustered patterns.

Figure 7. Univariate global Moran index of land urbanization and carbon emissions in the Yangtze River Delta urban agglomeration, 2010–2019.

4.4.2. Local Spatial Autocorrelation

To further examine the localized spatial associations of CCD, a local spatial autocorrelation analysis was conducted using GeoDa software 1.20 (GeoDa Center, Tempe, AZ, USA) (Figure 8). From 2010 to 2019, most areas in the Yangtze River Delta urban agglomeration exhibited no significant clustering characteristics in CCD. However, localized spatial aggregation patterns showed minor changes: (1) 2010: Spatial clusters emerged in the western region, with Wuhu forming a "high–high" cluster and Tongling, Anqing, and Xuancheng classified as "low–high" clusters. (2) 2013: Clustering diminished in the west, with only Chizhou remaining a "low–high" cluster. Shaoxing became a "low–low" cluster, while Changzhou emerged as a "high–high" cluster. Zhoushan appeared as a "high–low" outlier. (3) 2016: Most areas showed insignificant clustering, except Changzhou, which transitioned from "high–high" to "low–high". (4) 2019: Localized clusters persisted in Zhoushan ("low–high") and Xuancheng and Taizhou ("high–low"), but no widespread spatial aggregation patterns formed. These results highlight the dynamic and fragmented nature of spatial coordination patterns in the Yangtze River Delta urban agglomeration.

Specifically, in 2010, spatial clustering occurred in the western Yangtze River Delta urban agglomeration, where cities exhibited synergistic coordination. Wuhu formed a "high–high" cluster, indicating that its rising CCD level positively influenced neighboring cities. Tongling, Anqing, and Xuancheng were categorized as "low–high" clusters. By 2013, western cities exited clustering patterns, halting the synergistic development trend, with only Chizhou emerging as a "low–high" cluster. Simultaneously, Shaoxing became a "low–low" cluster, reflecting lagging development in land urbanization and energy consumption carbon emissions. Changzhou formed a "high–high" cluster, exerting a positive driving effect on coordination levels in surrounding cities. Zhoushan appeared as a "high–low" cluster, suggesting that its improved CCD level suppressed coordination in adjacent areas. In 2016, only Changzhou transitioned from a "high–high" to a "low–high" cluster, while other regions showed no significant clustering. This indicates that the CCD levels of land urbanization and energy consumption carbon emissions failed to form widespread spatial aggregation during this period, with limited interdependencies or driving effects among cities. By 2019, coordinated development remained localized rather than regionally aggregated. Zhoushan persisted as a "low–high" cluster, while Xuancheng and Taizhou became "high–low" clusters. These findings underscore the fragmented and unstable spatial coordination patterns across the Yangtze River Delta urban agglomeration.

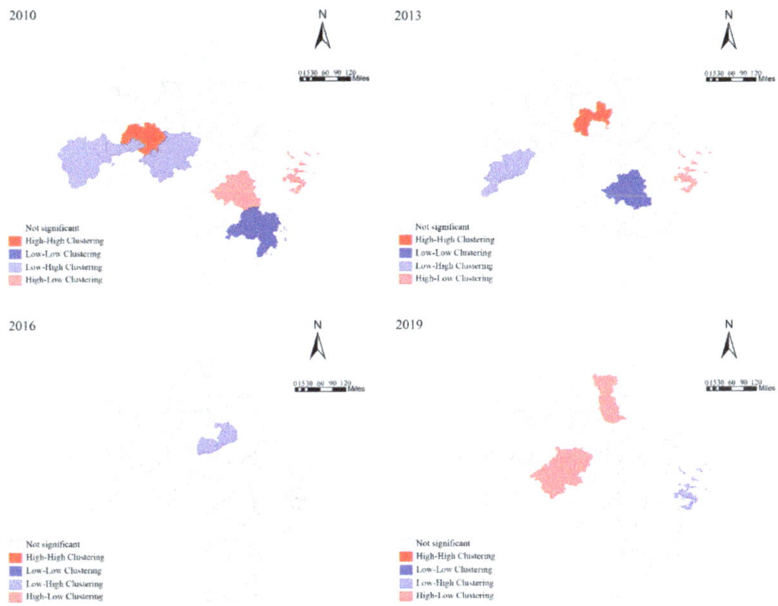

Figure 8. LISA aggregation diagram of the CCD between land urbanization and carbon emissions in the Yangtze River Delta urban agglomeration.

5. Discussion

5.1. Composite Index Characteristics of Land Urbanization and Carbon Emissions from Energy Consumption

This study constructs composite indices for land urbanization and carbon emissions from energy consumption, revealing the dynamic relationship between the two in the Yangtze River Delta urban agglomeration. From 2010 to 2019, the composite index for land urbanization increased from 67.471 to 128.804, while the composite index for carbon emissions decreased from 18.971 to 15.36, indicating a gradual coordination between urbanization and carbon emissions reduction. These findings are consistent with the studies of Xu et al. [15], Liu et al. [62] and Xu et al. [66] which also suggest that, as land urbanization progresses, carbon emissions are to some extent controlled in certain regions. However, this study further reveals that, despite overall improvements in coordination, regional differences remain significant. For instance, the composite index for land urbanization in Shanghai increased from 0.65 to 0.85, while in peripheral cities such as Suqian and Yangzhou, the index only increased from 0.40 to 0.52, indicating a disparity in low-carbon urbanization processes between core cities and peripheral cities [67,68].

These findings align with the research by Lin et al. [19] and Shan et al. [51], who also noted that improvements in land urbanization are generally associated with reductions in carbon emissions, but the actual effects are influenced by factors such as urban development stages and policy support. Therefore, this study emphasizes that, although core cities have made significant progress in low-carbon transformation, peripheral cities still face greater transformation pressures and require differentiated low-carbon policies.

5.2. Characteristics of the Horizontal Time Dynamics of the CCD

Based on the analysis of the Coupling Coordination Degree (CCD), this study finds that, from 2010 to 2019, the CCD for the Yangtze River Delta urban agglomeration steadily increased from 0.195 (severe imbalance) to 0.541 (near coordination). This upward trend aligns with the findings of Shan et al. [48], who also observed a gradual increase in the coordination between land urbanization and carbon emissions over time. However, despite the overall improvement in coordination, the CCD in peripheral cities remains lagging. For instance, the CCD for Shanghai increased from 0.75 to 0.90, while other cities such as Nanjing and Suzhou only saw an increase from 0.55 to 0.70, indicating slower low-carbon transformation in these cities.

Further analysis reveals that the coordination of the land urbanization subsystem steadily increased from 0.1 (severe imbalance) to 0.995 (high coordination), while the coordination of the carbon emission subsystem decreased from 0.902 (high coordination) to 0.296 (moderate imbalance). This asymmetric dynamic change suggests that carbon emission control lags behind rapid developments in land urbanization in the low-carbon urbanization process. These findings are consistent with Liu et al. [62], Pu et al. [55] and Hu et al. [69] who also pointed out that the improvement in coordination between land urbanization and carbon emissions is influenced by factors such as city characteristics and policy effects.

5.3. Characteristics of the Horizontal Spatial Dynamics of the CCD

Regarding horizontal spatial dynamics, this study further reveals the spatial heterogeneity between land urbanization and carbon emissions in the Yangtze River Delta urban agglomeration using a spatial autocorrelation analysis (Moran's I). The overall Moran's I value for the region from 2010 to 2019 was 0.42, indicating some degree of spatial clustering. However, there were significant differences in coordination among cities. For example, the Moran's I value for Shanghai was 0.76, indicating strong spatial clustering, while peripheral cities such as Zhenjiang and Huzhou had much lower Moran's I values of 0.32 and 0.28, respectively, suggesting that these cities lacked effective spatial coordination between land urbanization and carbon emissions.

This finding aligns with the research by Chen et al. [47], Zhu et al. [41] and Lv et al. [70], who also observed significant spatial imbalances within urban agglomerations, even though core areas exhibited higher coordination. Our study further emphasizes the role of spatial heterogeneity in the coordination between land urbanization and carbon emissions, particularly in peripheral cities [71]. These cities, due to lagging infrastructure and limited green technology, face difficulties in effectively controlling carbon emissions. Therefore, this study recommends enhancing policy support for these peripheral cities, promoting green technology innovation, and accelerating their low-carbon transformation to improve overall green transition levels in the urban agglomeration.

6. Conclusions

Over the past four decades of reform and liberalization, China's urbanization has progressed rapidly, transitioning through early- and mid-stage growth phases. However, this rapid urbanization has been accompanied by environmental degradation and rising carbon emissions. Addressing these dual challenges, exploring scientific pathways for the Yangtze River Delta to achieve high-quality urbanization and realize the national goals of "carbon peaking and carbon neutrality" holds significant importance for China's sustainable development. Based on panel data from 27 cities in the Yangtze River Delta urban agglomeration (2010–2019), this study constructs an evaluation system for land urbanization and energy consumption carbon emissions, analyzes the comprehensive indices of the system and its subsystems, and employs CCD models and spatial autocorrelation methods to examine the CCD and spatial correlations of the urban agglomeration across temporal and spatial dimensions. The main conclusions are as follows:

(1) The comprehensive level of the land urbanization and energy consumption carbon emission system in the Yangtze River Delta urban agglomeration exhibited a relatively stable upward trend (5.772%), with higher comprehensive indices in the southern region (85.635) and lower values in the northern region (61.912). The comprehensive level of the land urbanization subsystem continued to rise (4.762%), with higher indices in the southern and northern regions (67.396 and 54.672, respectively) and lower indices in the central region (37.126). The energy consumption carbon emission subsystem showed a continuous and stable decline (1.903%). The number of cities with low carbon emission levels increased significantly, rising from one city in 2010 to twenty cities in 2019. Conversely, cities with high carbon emission levels decreased substantially, dropping from twenty-six cities in 2010 to seven cities in 2019.
(2) The CCD between land urbanization and energy consumption carbon emissions in the Yangtze River Delta urban agglomeration exhibited an overall upward trend, reaching the marginal coordination stage (0.541) after a decade of development. Significant regional disparities in CCD were observed: the southeastern region outperformed the northwestern region, with the eastern coastal areas maintaining consistently high CCD levels over time. By 2019, all 27 cities had exited imbalance stages (CCD $\geq$ 0.528), indicating improved intercity connectivity and coordination. However, most cities had not yet achieved good coordination or high-quality coordination stages. Only four cities—Shanghai, Huzhou, Taizhou, and Wenzhou—reached the high-quality coordination stage (CCD $\geq$ 0.907). Additionally, the CCD gap between cities has narrowed over time, with higher CCD values gradually expanding from peripheral areas toward central regions.
(3) The horizontal spatial correlation relationship of the CCD between land urbanization and energy consumption carbon emissions in the Yangtze River Delta urban agglomeration remained unstable. In 2013, the Global Moran's I value was −1.02, while in other years, it fluctuated around 0.109, reflecting regional disparities in CCD levels. Most areas exhibited no significant clustering characteristics, and local spatial clustering patterns exhibited variations over time.

Integrating the principles of the 'Outline of the Yangtze River Delta Regional Integration Development Plan' and the 'Yangtze River Delta Urban Agglomeration Development Plan', and based on the research findings, the following recommendations are proposed to address the challenges in coordinating land urbanization and energy consumption carbon emissions and to promote sustainable development in the Yangtze River Delta urban agglomeration:

(1) Establish a regional collaborative mechanism for the development of land urbanization and energy consumption carbon emissions, fully leveraging the driving role of key cities. Support the four cities in the high-quality coordination stage—Shanghai, Huzhou, Taizhou, and Wenzhou—to break administrative barriers restricting urban agglomeration development and strengthen their leading role.
(2) Enhance macro-control to promote coordinated development of regional land urbanization and energy consumption carbon emissions, improve their own development levels, and amplify their driving and coordinating effects, thereby guiding the integrated development of the Yangtze River Delta. Key cities should strengthen cooperation with other cities and share resources. Emphasize the interconnectivity between cities.

(3) Enhancing the high-quality development of central cities will contribute to the overall high-quality development of the Yangtze River Delta urban agglomeration and advance low-carbon urbanization.
(4) Optimize industrial structures and promote a green, low-carbon circular economy system. Cities should facilitate dynamic transformation and green industrial restructuring, accelerating the shift from secondary industries to tertiary industries.
(5) Given the significant spatial disparities in the Yangtze River Delta, policymakers must fully consider regional differences and avoid "one-size-fits-all" approaches in policy formulation.
(6) Land urbanization is heavily influenced by national macro-control, policy and institutional reforms, and local government actions. Policymakers should formulate territorial spatial planning policies under the "carbon peaking and carbon neutrality" goals, considering the relationship between urbanization stages and the scale, structure, and layout of territorial spaces with carbon sinks and emissions, thereby ensuring low-carbon urban development and the achievement of green and low-carbon transitions.

The empirical results of this study can guide China and other developing countries in pursuing coordinated development between land urbanization and energy consumption carbon emissions, aiding the formulation of effective interventions around urbanization and environmental pollution. Inevitably, this study has certain limitations. First, the selection of system indicators was constrained by data availability and may not fully represent the development of land urbanization and carbon emissions in the Yangtze River Delta urban agglomeration. Future research could incorporate additional indicators related to non-fossil energy use and production-related carbon emissions. Second, while this study identifies key indicators of the land urbanization and carbon emission systems, it does not delve into their interrelationships. Future studies could employ methods such as linear regression models to further explore the correlations and impact magnitudes among these indicators. Third, there is no unified standard for classifying CCD intervals. This study adopts a ten-tier classification based on CCD data and the existing literature, which may introduce inaccuracies. Future research should flexibly select alternative classification criteria aligned with regional development realities. Finally, fieldwork and data collection were partially affected by factors such as the COVID-19 pandemic. This study selected 27 cities in China's Yangtze River Delta urban agglomeration as case studies. Future research could investigate the CCD and spatial correlations between land urbanization and energy consumption carbon emissions across China's 34 provincial-level administrative regions. Such an approach would reveal more comprehensive and distinct characteristics, provide stronger empirical cases, and offer guidance for planning, management, and decision-making.

Author Contributions: Conceptualization, Z.L.; Methodology, Z.L.; Software, Z.L.; Validation, Z.L.; Formal analysis, Z.L., X.Z. and T.L.; Investigation, B.L.; Resources, Y.R.; Data curation, Z.L.; Writing—original draft, Z.L.; Writing—review & editing, Y.Y.; Visualization, Z.L. and N.S. All authors have read and agreed to the published version of the manuscript.

Funding: The financial support is provided by the Low Carbon Model and Planning Design Optimization Technology for New Urban Areas (2018YFC0704604) (Topic 4).

Data Availability Statement: The original contributions presented in the study are included in the article, further inquiries can be directed to the corresponding author.

Conflicts of Interest: The authors declare no conflicts of interest.

References

1. IPCC. Summary for Policymakers. In *Global Warming of 1.5 °C. An IPCC Special Report on the Impacts of Global Warming of 1.5 °C Above Pre-Industrial Levels and Related Global Greenhouse Gas Emission Pathways, in the Context of Strengthening the Global Response to the Threat of Climate Change, Sustainable Development, and Efforts to Eradicate Poverty*; Masson-Delmotte, V., Zhai, P., Pörtner, H.-O., Roberts, D., Skea, J., Shukla, P.R., Pirani, A., Moufouma-Okia, W., Péan, C., Pidcock, R., et al., Eds.; Cambridge University Press: Cambridge, UK; New York, NY, USA, 2022; pp. 3–24. [CrossRef]
2. Yin, H.; Xiao, R.; Fei, X.; Zhang, Z.; Gao, Z.; Wan, Y.; Tan, W.; Jiang, X.; Cao, W.; Guo, Y. Analyzing "Economy-Society-Environment" Sustainability from the Perspective of Urban Spatial Structure: A Case Study of the Yangtze River Delta Urban Agglomeration. *Sustain. Cities Soc.* **2023**, *96*, 104691. [CrossRef]
3. Ahrend, R.; Farchy, E.; Kaplanis, I.; Lembcke, A.C. *What Makes Cities More Productive? Evidence on the Role of Urban Governance from Five OECD Countries*; OECD Publishing: Paris, France, 2014. [CrossRef]
4. UN-Habitat. *World Cities Report 2022: Envisaging the Future of Cities*; United Nations Human Settlements Programme: Nairobi, Kenya, 2022.
5. IPCC. Summary for Policymakers. In *Climate Change 2022: Mitigation of Climate Change. Contribution of Working Group III to the Sixth Assessment Report of the Intergovernmental Panel on Climate Change*; Skea, J., Shukla, P.R., Reisinger, A., Slade, R.B., Pathak, M., Al Khourdajie, A., van Diemen, R., Abdulla, A., Akimoto, K., Babiker, M., et al., Eds.; Cambridge University Press: Cambridge, UK; New York, NY, USA, 2022. [CrossRef]
6. Zhang, S.; Miao, X.; Zheng, H.; Chen, W.; Wang, H. Spatial Functional Division in Urban Agglomerations and Carbon Emission Intensity: New Evidence from 19 Urban Agglomerations in China. *Energy* **2024**, *300*, 131541. [CrossRef]
7. Dogan, E.; Turkekul, B. CO_2 Emissions, Real Output, Energy Consumption, Trade, Urbanization and Financial Development: Testing the EKC Hypothesis for the USA. *Environ. Sci. Pollut. Res.* **2016**, *23*, 1203–1213. [CrossRef]
8. Azam, M.; Uddin, I.; Khan, S.; Tariq, M. Are Globalization, Urbanization, and Energy Consumption Cause Carbon Emissions in SAARC Region? New Evidence from CS-ARDL Approach. *Environ. Sci. Pollut. Res.* **2022**, *29*, 87746–87763. [CrossRef]
9. Carpio, A.; Ponce-Lopez, R.; Lozano-García, D.F. Urban Form, Land Use, and Cover Change and Their Impact on Carbon Emissions in the Monterrey Metropolitan Area, Mexico. *Urban Clim.* **2021**, *39*, 100947. [CrossRef]
10. Alsaggaf, M.I. Exploring the Role of Land Utilization, Renewable Energy, and ICT to Counter the Environmental Emission: A Panel Study of Selected G20 and OECD Countries. *Land Degrad. Dev.* **2025**, *36*, 1707–1723. [CrossRef]
11. Jo, H.; Kim, H. The Influence of Urban Spatial Structure on Building Carbon Emissions at the Neighborhood Scale Considering Spatial Effect. *Build. Environ.* **2025**, *276*, 112888. [CrossRef]
12. Chandra Voumik, L.; Sultana, T. Impact of Urbanization, Industrialization, Electrification and Renewable Energy on the Environment in BRICS: Fresh Evidence from Novel CS-ARDL Model. *Heliyon* **2022**, *8*, e11457. [CrossRef]
13. Tanveer, A.; Song, H.; Faheem, M.; Daud, A. Caring for the Environment. How Do Deforestation, Agricultural Land, and Urbanization Degrade the Environment? Fresh Insight through the ARDL Approach. *Environ. Dev. Sustain.* **2024**, *27*, 11527–11562. [CrossRef]
14. Zhu, P.; Ahmed, Z.; Pata, U.K.; Khan, S.; Abbas, S. Analyzing Economic Growth, Eco-Innovation, and Ecological Quality Nexus in E-7 Countries: Accounting for Non-Linear Impacts of Urbanization by Using a New Measure of Ecological Quality. *Environ. Sci. Pollut. Res.* **2023**, *30*, 94242–94254. [CrossRef]
15. Xu, H.; Zhang, W. The Causal Relationship between Carbon Emissions and Land Urbanization Quality: A Panel Data Analysis for Chinese Provinces. *J. Clean. Prod.* **2016**, *137*, 241–248. [CrossRef]
16. Tang, M.; Hu, F. Land Urbanization and Urban CO_2 Emissions: Empirical Evidence from Chinese Prefecture-Level Cities. *Heliyon* **2023**, *9*, e19834. [CrossRef] [PubMed]
17. Li, X. Local Government Decision-Making Competition and Regional Carbon Emissions: Experience Evidence and Emission Reduction Measures. *Sustain. Energy Technol. Assess.* **2022**, *50*, 101800. [CrossRef]
18. Tang, M.; Hu, F. How Does Land Urbanization Promote CO_2 Emissions Reduction? *Evidence From Chinese Prefectural-Level Cities. Front. Environ. Sci.* **2021**, *9*, e19834. [CrossRef]
19. Lin, X.; Lu, C.; Song, K.; Su, Y.; Lei, Y.; Zhong, L.; Gao, Y. Analysis of Coupling Coordination Variance between Urbanization Quality and Eco-Environment Pressure: A Case Study of the West Taiwan Strait Urban Agglomeration, China. *Sustainability* **2020**, *12*, 2643. [CrossRef]
20. Huang, W.; Li, J. The Coupling Relationship Between Urbanization and Carbon Emissions from Land Use in Ningxia. *Front. Environ. Sci.* **2022**, *10*, 927798. [CrossRef]
21. Jiang, J.; Zhu, S.; Wang, W.; Li, Y.; Li, N. Coupling Coordination between New Urbanisation and Carbon Emissions in China. *Sci. Total Environ.* **2022**, *850*, 158076. [CrossRef]
22. Zhang, D.; Wang, Z.; Li, S.; Zhang, H. Impact of Land Urbanization on Carbon Emissions in Urban Agglomerations of the Middle Reaches of the Yangtze River. *Int. J. Environ. Res. Public Health* **2021**, *18*, 1403. [CrossRef]

23. Bai, Y.; Deng, X.; Jiang, S.; Zhang, Q.; Wang, Z. Exploring the Relationship between Urbanization and Urban Eco-Efficiency: Evidence from Prefecture-Level Cities in China. *J. Clean. Prod.* **2018**, *195*, 1487–1496. [CrossRef]
24. Ahmad, M.; Akram, W.; Ikram, M.; Shah, A.A.; Rehman, A.; Chandio, A.A.; Jabeen, G. Estimating Dynamic Interactive Linkages among Urban Agglomeration, Economic Performance, Carbon Emissions, and Health Expenditures across Developmental Disparities. *Sustain. Prod. Consum.* **2021**, *26*, 239–255. [CrossRef]
25. Rehman, A.; Ma, H.; Radulescu, M.; Sinisi, C.I.; Paunescu, L.M.; Alam, M.S.; Alvarado, R. The Energy Mix Dilemma and Environmental Sustainability: Interaction among Greenhouse Gas Emissions, Nuclear Energy, Urban Agglomeration, and Economic Growth. *Energies* **2021**, *14*, 7703. [CrossRef]
26. Navamuel, E.L.; Rubiera Morollón, F.; Moreno Cuartas, B. Energy Consumption and Urban Sprawl: Evidence for the Spanish Case. *J. Clean. Prod.* **2018**, *172*, 3479–3486. [CrossRef]
27. Loredana, C.; Rehman, A.; Maria Mirabela, F.I.; Pinzon, S.; Cismaș, L.M. What Implications Do Primary Energy Use, Urban Population Agglomeration, and Economic Development Rendered to Romania's Environmental Sustainability? *Energy Strategy Rev.* **2024**, *53*, 101399. [CrossRef]
28. Jarboui, S.; Bouzouina, L.; Alofaysan, H. Historical Insights into CO_2 Emission Dynamics in Urban Daily Mobility: A Case Study of Lyon's Agglomeration. *Sustainability* **2024**, *16*, 9789. [CrossRef]
29. Frolking, S.; Milliman, T.; Seto, K.C.; Friedl, M.A. A Global Fingerprint of Macro-Scale Changes in Urban Structure from 1999 to 2009. *Environ. Res. Lett.* **2013**, *8*, 24004. [CrossRef]
30. Paravantis, J.A.; Tasios, P.D.; Dourmas, V.; Andreakos, G.; Velaoras, K.; Kontoulis, N.; Mihalakakou, P. A Regression Analysis of the Carbon Footprint of Megacities. *Sustainability* **2021**, *13*, 1379. [CrossRef]
31. Rehman, A.; Ma, H.; Ozturk, I.; Alvarado, R.; Oláh, J.; Liu, R.; Qiang, W. The Enigma of Environmental Sustainability and Carbonization: Assessing the Connection between Coal and Oil Rents, Natural Resources, and Environmental Quality. *Gondwana Res.* **2024**, *128*, 1–13. [CrossRef]
32. Talib, M.N.A.; Hashmi, S.H.; Aamir, M.; Khan, M.A. Testing Non-Linear Effect of Urbanization on Environmental Degradation: Cross-Country Evidence. *Front. Environ. Sci.* **2022**, *10*, 971394. [CrossRef]
33. Darwish, A.M.; Zagow, M.; Elkafoury, A. Impact of Land Use, Travel Behavior, and Socio-Economic Characteristics on Carbon Emissions in Cool-Climate Cities, USA. *Environ. Sci. Pollut. Res.* **2023**, *30*, 91108–91124. [CrossRef]
34. Liu, K.; Xue, M.; Peng, M.; Wang, C. Impact of Spatial Structure of Urban Agglomeration on Carbon Emissions: An Analysis of the Shandong Peninsula, China. *Technol. Forecast. Soc. Chang.* **2020**, *161*, 120313. [CrossRef]
35. Dong, Y.; Jin, G.; Deng, X. Dynamic Interactive Effects of Urban Land-Use Efficiency, Industrial Transformation, and Carbon Emissions. *J. Clean. Prod.* **2020**, *270*, 122547. [CrossRef]
36. Zhou, Y.; Chen, M.; Tang, Z.; Mei, Z. Urbanization, Land Use Change, and Carbon Emissions: Quantitative Assessments for City-Level Carbon Emissions in Beijing-Tianjin-Hebei Region. *Sustain. Cities Soc.* **2021**, *66*, 102701. [CrossRef]
37. Yu, Q.; Li, M.; Li, Q.; Wang, Y.; Chen, W. Economic Agglomeration and Emissions Reduction: Does High Agglomeration in China's Urban Clusters Lead to Higher Carbon Intensity? *Urban Clim.* **2022**, *43*, 101174. [CrossRef]
38. Yang, L.; Meng, H.; Wang, J.; Wu, Y.; Zhao, Z. The Vulnerability Assessment and Obstacle Factor Analysis of Urban Agglomeration along the Yellow River in China from the Perspective of Production-Living-Ecological Space. *PLoS ONE* **2024**, *19*, e0299729. [CrossRef]
39. Xu, Q.; Dong, Y.; Yang, R. Urbanization Impact on Carbon Emissions in the Pearl River Delta Region: Kuznets Curve Relationships. *J. Clean. Prod.* **2018**, *180*, 514–523. [CrossRef]
40. Chen, J.; Wang, L.; Li, Y. Research on the Impact of Multi-Dimensional Urbanization on China's Carbon Emissions under the Background of COP21. *J. Environ. Manag.* **2020**, *273*, 111123. [CrossRef]
41. Zhu, E.; Qi, Q.; Chen, L.; Wu, X. The Spatial-Temporal Patterns and Multiple Driving Mechanisms of Carbon Emissions in the Process of Urbanization: A Case Study in Zhejiang, China. *J. Clean. Prod.* **2022**, *358*, 131954. [CrossRef]
42. Fan, J.; Zhou, L. Impact of Urbanization and Real Estate Investment on Carbon Emissions: Evidence from China's Provincial Regions. *J. Clean. Prod.* **2019**, *209*, 309–323. [CrossRef]
43. Liu, Y.; Yan, B.; Zhou, Y. Urbanization, Economic Growth, and Carbon Dioxide Emissions in China: A Panel Cointegration and Causality Analysis. *J. Geogr. Sci.* **2016**, *26*, 131–152. [CrossRef]
44. Dong, L.; Longwu, L.; Zhenbo, W.; Liangkan, C.; Faming, Z. Exploration of Coupling Effects in the Economy–Society–Environment System in Urban Areas: Case Study of the Yangtze River Delta Urban Agglomeration. *Ecol. Indic.* **2021**, *128*, 107858. [CrossRef]
45. Wang, Y.; Niu, Y.; Li, M.; Yu, Q.; Chen, W. Spatial Structure and Carbon Emission of Urban Agglomerations: Spatiotemporal Characteristics and Driving Forces. *Sustain. Cities Soc.* **2022**, *78*, 103600. [CrossRef]
46. Xing, L.; Xue, M.; Hu, M. Dynamic Simulation and Assessment of the Coupling Coordination Degree of the Economy–Resource–Environment System: Case of Wuhan City in China. *J. Environ. Manag.* **2019**, *230*, 474–487. [CrossRef]
47. Chen, H.; Hua, Y.; Xu, Y. Spatial-Temporal Evolution Patterns and Obstacle Factors of Urban–Rural "Economy–Society–Ecology" Coordination in the Yangtze River Delta. *Sustainability* **2023**, *15*, 13839. [CrossRef]

48. Shan, Y.; Liu, J.; Liu, Z.; Shao, S.; Guan, D. An Emissions-Socioeconomic Inventory of Chinese Cities. *Sci. Data* **2019**, *6*, 190027. [CrossRef]
49. Shan, Y.; Guan, D.; Liu, J.; Mi, Z.; Liu, Z.; Liu, J.; Schroeder, H.; Cai, B.; Chen, Y.; Shao, S.; et al. Methodology and Applications of City Level CO_2 Emission Accounts in China. *J. Clean. Prod.* **2017**, *161*, 1215–1225. [CrossRef]
50. Shan, Y.; Guan, D.; Hubacek, K.; Zheng, B.; Davis, S.J.; Jia, L.; Liu, J.; Liu, Z.; Fromer, N.; Mi, Z.; et al. City-Level Climate Change Mitigation in China. *Sci. Adv.* **2018**, *4*, eaaq0390. [CrossRef]
51. Shan, Y.; Guan, Y.; Hang, Y.; Zheng, H.; Li, Y.; Guan, D.; Li, J.; Zhou, Y.; Li, L.; Hubacek, K. City-Level Emission Peak and Drivers in China. *Sci. Bull.* **2022**, *67*, 1910–1920. [CrossRef]
52. Lin, X.; Wang, Y.; Wang, S.; Wang, D. Spatial Differences and Driving Forces of Land Urbanization in China. *J. Geogr. Sci.* **2015**, *25*, 545–558. [CrossRef]
53. Li, Y.; Li, Y.; Zhou, Y.; Shi, Y.; Zhu, X. Investigation of a Coupling Model of Coordination between Urbanization and the Environment. *J. Environ. Manag.* **2012**, *98*, 127–133. [CrossRef]
54. Zhang, G.; Zhang, N.; Liao, W. How Do Population and Land Urbanization Affect CO2 Emissions under Gravity Center Change? A Spatial Econometric Analysis. *J. Clean. Prod.* **2018**, *202*, 510–523. [CrossRef]
55. Pu, Y.; Wang, Y.; Wang, P. Driving Effects of Urbanization on City-Level Carbon Dioxide Emissions: From Multiple Perspectives of Urbanization. *Int. J. Urban Sci.* **2022**, *26*, 108–128. [CrossRef]
56. Ma, L.; Xiang, L.; Wang, C.; Chen, N.; Wang, W. Spatiotemporal Evolution of Urban Carbon Balance and Its Response to New-Type Urbanization: A Case of the Middle Reaches of the Yangtze River Urban Agglomerations, China. *J. Clean. Prod.* **2022**, *380*, 135122. [CrossRef]
57. Zhang, W.; Xu, H. Effects of Land Urbanization and Land Finance on Carbon Emissions: A Panel Data Analysis for Chinese Provinces. *Land Use Policy* **2017**, *63*, 493–500. [CrossRef]
58. Shannon, C.E. A Mathematical Theory of Communication. *Bell Syst. Tech. J.* **1948**, *27*, 379–423. [CrossRef]
59. Shemshadi, A.; Shirazi, H.; Toreihi, M.; Tarokh, M.J. A Fuzzy VIKOR Method for Supplier Selection Based on Entropy Measure for Objective Weighting. *Expert Syst. Appl.* **2011**, *38*, 12160–12167. [CrossRef]
60. Li, D.; Cao, L.; Zhou, Z.; Zhao, K.; Du, Z.; Han, K. Coupling Coordination Degree and Driving Factors of New-Type Urbanization and Low-Carbon Development in the Yangtze River Delta: Based on Nighttime Light Data. *Environ. Sci. Pollut. Res.* **2022**, *29*, 81636–81657. [CrossRef]
61. Mazziotta, M.; Pareto, A. Synthesis of Indicators: The Composite Indicators Approach. In *Complexity in Society: From Indicators Construction to Their Synthesis*; Maggino, F., Ed.; Springer: Cham, Switzerland, 2017; Volume 70, pp. 159–191.
62. Liu, C.; Sun, W.; Li, P. Characteristics of Spatiotemporal Variations in Coupling Coordination between Integrated Carbon Emission and Sequestration Index: A Case Study of the Yangtze River Delta, China. *Ecol. Indic.* **2022**, *135*, 108520. [CrossRef]
63. Fu, S.; Zhuo, H.; Song, H.; Wang, J.; Ren, L. Examination of a Coupling Coordination Relationship between Urbanization and the Eco-Environment: A Case Study in Qingdao, China. *Environ. Sci. Pollut. Res.* **2020**, *27*, 23981–23993. [CrossRef] [PubMed]
64. Schandl, H.; Hatfield-Dodds, S.; Wiedmann, T.; Geschke, A.; Cai, Y.; West, J.; Newth, D.; Baynes, T.; Lenzen, M.; Owen, A. Decoupling Global Environmental Pressure and Economic Growth: Scenarios for Energy Use, Materials Use and Carbon Emissions. *J. Clean. Prod.* **2016**, *132*, 45–56. [CrossRef]
65. Wrigley, N.; Cliff, A.D.; Ord, J.K. Spatial Processes: Models and Applications. *Geogr. J.* **1982**, *148*, 383. [CrossRef]
66. Xu, H.; Jiao, M. City Size, Industrial Structure and Urbanization Quality—A Case Study of the Yangtze River Delta Urban Agglomeration in China. *Land Use Policy* **2021**, *111*, 105735. [CrossRef]
67. Yang, S.; Wang, M.Y.; Wang, C. Revisiting and Rethinking Regional Urbanization in Changjiang River Delta, China. *Chin. Geogr. Sci.* **2012**, *22*, 617–625. [CrossRef]
68. Yang, Q.; Ding, L.; Wang, L.; Liu, C.; Fan, Y.; Li, Y.; Wang, Y. Influence Mechanism of New-Type Urbanization on Urban Land Use Efficiency in the Yangtze River Delta, China. *Chin. Geogr. Sci.* **2023**, *33*, 474–488. [CrossRef]
69. Hu, C.; Liu, S.; Wang, Y.; Zhang, M.; Xiao, W.; Wang, W.; Xu, J. Anthropogenic CO_2 Emissions from a Megacity in the Yangtze River Delta of China. *Environ. Sci. Pollut. Res.* **2018**, *25*, 23157–23169. [CrossRef] [PubMed]
70. Lv, T.; Hu, H.; Zhang, X.; Xie, H.; Wang, L.; Fu, S. Spatial Spillover Effects of Urbanization on Carbon Emissions in the Yangtze River Delta Urban Agglomeration, China. *Environ. Sci. Pollut. Res.* **2022**, *29*, 33920–33934. [CrossRef]
71. Zhao, Y.; Wang, S.; Zhou, C. Understanding the Relation between Urbanization and the Eco-Environment in China's Yangtze River Delta Using an Improved EKC Model and Coupling Analysis. *Sci. Total Environ.* **2016**, *571*, 862–875. [CrossRef]

buildings

MDPI

Article

Optimization Design Research of Architectural Layout and Morphology in Multi-Story Dormitory Areas Based on Wind Environment Analysis

Xiangru Chen, Haoran Kang, Juanru Zhao and Qibo Liu *

School of Architecture, Chang'an University, Xi'an 710061, China; 2021905921@chd.edu.cn (X.C.); 2024241025@chd.edu.cn (H.K.); 2023241007l5@chd.edu.cn (J.Z.)
* Correspondence: lucy@chd.edu.cn; Tel.: +86-135-7250-8860

Abstract: Optimizing the wind environment within university dormitory areas is essential for ensuring student safety, enhancing living comfort, and improving building energy efficiency. In this study, the wind environment of multi-story university dormitories in cold regions is comprehensively investigated through computational fluid dynamics (CFD) simulations conducted with the PHONECIS software (version 2019), combined with orthogonal experimental design methods for systematic analysis and optimization. Through orthogonal experimental design, the effects of key morphological parameters—including building layout, length, width, and height—on the near-ground wind environment were evaluated. Among these, building width exerted the greatest influence, followed by building length, layout form, and finally building height. Based on the analysis, the optimal design scheme features a staggered building layout, with individual dormitory buildings measuring 60 m in length, 16 m in width, and 11.4 m in height. This optimized design was implemented in the multi-story dormitory area of the eastern section of Chang'an University's New Campus. A comparative analysis of wind speed distribution before and after optimization, conducted specifically for the outdoor spaces during the winter season, revealed that the average near-ground wind speed was reduced from 3.3 m/s to 2.7 m/s, achieving an 18% reduction. The staggered arrangement and adjusted building proportions effectively dispersed airflow, mitigated high-velocity zones, and significantly enhanced outdoor wind comfort and pedestrian safety. This study introduces a morphology–wind environment coupling strategy from an architectural perspective to guide the design of dormitory buildings in cold regions. Rather than focusing on mathematical modeling, the research emphasizes design-oriented outcomes aimed at informing and optimizing practical architectural solutions for safer, more comfortable, and energy-efficient campus living environments.

Keywords: multi-story dormitory area; outdoor wind environment; morphological layout; CFD simulation; optimization design

1. Introduction and Literature Review

1.1. Introduction

In Chinese universities, students are typically accommodated in centralized multi-story dormitory complexes, which are integral components of the overall campus planning. Among these, multi-story dormitory buildings constitute a significant portion [1]. A favorable outdoor wind environment is crucial for the comfort and well-being of students [2]. It not only enhances their quality of life but also plays a key role in the energy efficiency and safety of multi-story dormitory buildings [3]. Therefore, optimizing the outdoor wind environment is a key factor in achieving sustainable campus planning.

Although significant progress has been made in optimizing the wind environment of buildings, research specifically focused on the wind conditions within university multi-story dormitory areas remains relatively scarce. Most existing studies, on the one hand, concentrate on the wind environments of high-rise towers or urban street canyons [4], with limited attention given to campus dormitory complexes. On the other hand, many studies on campus dormitories tend to examine the effects of individual factors, often lacking a systematic approach for the comprehensive optimization of multiple influencing factors [5–9]. For example, while some studies propose localized improvements, such as enhancing greenery or optimizing the design of individual buildings, there is a noticeable absence of a comprehensive optimization framework or multi-scenario comparisons for the wind environment in dormitory areas [10].

In conclusion, there exists an academic gap in systematically studying the wind environment of multi-story dormitory complexes under varying climatic conditions, identifying key influencing factors, and proposing targeted optimization strategies. This gap presents both the entry point and motivation for the current study.

1.2. Literature Review

Numerous studies have examined the influence of building layout forms, dimensions, and shapes on pedestrian wind environments, providing theoretical support for building design (Table 1). Zheng et al. [11] used CFD and wind tunnel tests to evaluate residential layouts and surrounding building layers in cold regions, highlighting the role of height and density. Ma et al. [12] found that middle-level, non-enclosed layouts are optimal for low-density areas, while middle-high-rise forms suit higher-density zones. Key parameters such as coverage rate and frontal area density also significantly affect wind comfort.

With respect to building dimensions, Mou et al. [13] demonstrated that aspect ratio and building width strongly impact wind speed and pressure distribution, with taller, narrower buildings generating higher wind speeds but reduced comfort zones.

In terms of building shapes, Qureshi et al. [14] showed that L-shaped and U-shaped buildings enhance ventilation when appropriately spaced. While much of the literature centers on high-rise buildings, multi-story dormitory buildings in cold climates remain underexplored [15–17].

In building wind environment research, three primary methods are widely used: field measurements, wind tunnel testing, and computational fluid dynamics (CFD) simulations. Field measurements offer on-site wind data but are limited by site constraints and long durations and are mainly applicable to existing buildings [18]. Wind tunnel tests, while accurate in simulating wind flow under varying layouts, require costly equipment and strict experimental conditions, restricting their broader application [19]. In contrast, CFD simulations have become the preferred approach due to their efficiency, lower cost, and ability to generate comprehensive spatial datasets for detailed analysis and optimization [14,20–22]. In this study, a combination of field measurements and CFD simulation is adopted to leverage the strengths of both methods, improving the reliability and accuracy of the wind environment evaluation.

Table 1. Overview of foundational studies on building morphology and wind environment excluding dormitory contexts.

Research Factors	Building Types	Specific Factors	Literature	Research Methods
Layout forms	High-rise residential building	Influence factors of layout pattern on wind environment	[4,11,12,23]	Numerical simulation
		Influence of greening and building layout patterns on the wind and Thermal environment of residential quarters	[10,20,21]	Numerical simulation
		Influence factors of layout pattern on wind environment	[21]	Wind tunnel test
		Influence factors of layout pattern on wind environment	[4,11,14]	Numerical simulation + Wind tunnel test
		Influence factors of layout pattern on wind environment	[12,22]	Numerical simulation + On-site measurement
	Multistory residential building	Influence of layout patterns on outdoor wind environment	[4,12]	Numerical simulation
	Block buildings	Influence of layout mode on wind environment	[23,24]	Numerical simulation
	Block buildings	Influence of layout mode on wind environment	[19]	Numerical simulation + Wind tunnel test
	/	Influence of layout mode on wind environment	[25,26]	Numerical simulation
		Influence of layout patterns and plant arrangements on wind environment and thermal comfort	[27]	Numerical simulation
Building dimensions	Block buildings	Influence of building height and width on wind environment	[27–29]	Numerical simulation
	/	Influence of building height and width on wind environment	[13,30]	Numerical simulation
Building shapes	High-rise residential building	Influencing factors of building form on wind environment for surrounding pedestrians	[14]	Numerical simulation + Wind tunnel test
	Office building	The relationship between architectural form and wind environment	[18]	Numerical simulation
	/	The influence of high-rise building forms on the surrounding environment	[15–17]	Numerical simulation

Note: / Undefined building type, for testing ideal models.

2. Materials and Methods

2.1. Selection of Research Cases

Xi'an, located in Shaanxi Province, China, is classified as Zone IIB in the national building climate zoning for cold regions [31]. The city experiences distinct seasonal variations, with average temperatures ranging from −1.2 °C to 0.0 °C in January (the coldest month) and from 26.3 °C to 26.6 °C in July (the hottest month). Wind environment data for both winter and summer conditions in Xi'an are presented in Table 2 [32]. According to the Beaufort scale as defined in GB/T 19201-2006 [33], the prevailing wind force in Xi'an is categorized as a light breeze, indicating relatively low wind speeds throughout the year. Given these climatic characteristics, the planning and design of dormitory areas must address both wind protection during winter and adequate natural ventilation during the summer months [34].

Table 2. Wind direction, wind speed and wind frequency table of Xi'an City [35].

Season	Months	Prevailing Wind Direction	Average Wind Speed	Wind Frequency	Subdominant Wind Direction	Average Wind Speed	Wind Frequency
Summer	June, July, August	Northeast by east	2.5 m/s	42.2%	Southwest	2.3 m/s	18.1%
Winter	December, January, February	Northeast by east	2.2 m/s	37.5%	Southwest	2.0 m/s	25.1%

In addition, Xi'an is situated within a seismic intensity zone classified at magnitude 8. This imposes specific seismic fortification requirements that restrict architectural variability and influence building design. Consequently, dormitory complexes in the region are primarily composed of multi-story buildings arranged in regular, uniform layouts. This architectural typology reflects a key characteristic of dormitory design in seismic-prone, cold-climate regions.

This study uses the multi-story dormitory area in the typical slab form of Xi'an as a research case for field investigation. The measurement points are located in three distinct multi-story dormitory areas: the determinant formation in the eastern section of the new campus of CHD, the enclosed formation in the western section of the new campus of Xidian University (XD), and the staggered formation in the eastern section of the new campus of XD. An overview of these case buildings is provided in Table 3. These three cases are representative examples of multi-story dormitory layouts, chosen to capture a range of architectural configurations.

Table 3. Basic overview and measuring point distribution of typical cases.

Case 1. Multistory Dormitories in the East Area of the New Campus of CHD		
Orientation	15 degrees east of south	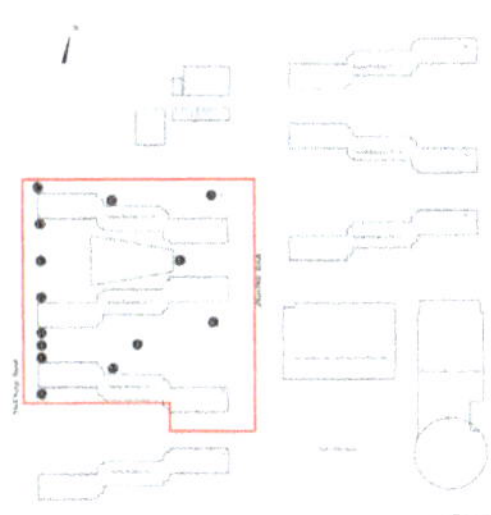 
Number of floors and height	Middle part: 6 floors. Both sides: 5 floors. 21.2 m.	
Architectural form	Plate multi-layer	
Layout form	Determinant type	(a) General layout and measured scheme (b) Facade real scene
Case 2. Multi-story dormitories in the west area of XD's new campus		
Orientation	30 degrees west of south	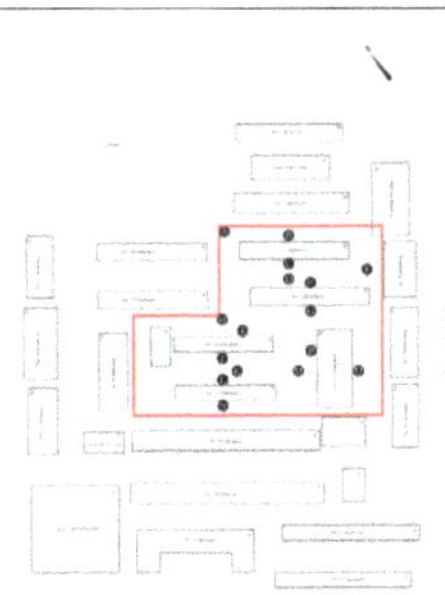 
Number of floors and height	6 layers, 22.5 m	
Architectural form	Plate-type multi-story	
Layout form	Enclosed type	(c) General layout and measured scheme (d) Facade real scene

Table 3. *Cont.*

Case 3. Multi-story dormitory in the east area of XD's new campus		
Orientation	30 degrees south by west.	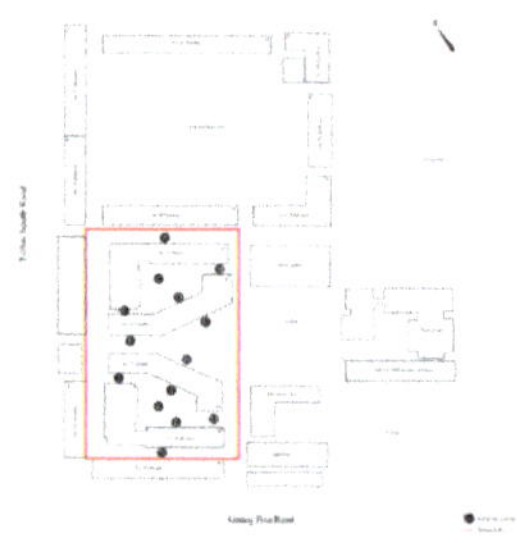  (e) General layout and measured scheme (f) Facade real scene
Number of floors and height	3 layers, 13.8 m	
Architectural form	Plate-type multi-story	
Layout form	Staggered type	

2.2. Numerical Simulation and Orthogonal Test Method

To analyze the relationship between the architectural layout of multi-story dormitory areas in colleges and universities and the outdoor wind environment, simulations are conducted on the selected research cases. This study employs computational fluid dynamics (CFD) numerical simulation and orthogonal experimental design, each offering significant advantages in the study of natural ventilation driven by building-induced wind. Both methods contribute to enhancing the efficiency of design and analysis. CFD numerical simulation provides accurate flow field data by simulating the gas flow dynamics both inside and outside the building. This overcomes the limitations of experimental conditions and scales inherent in traditional experiments. CFD can handle complex building geometries and offer detailed airflow analysis in varying wind environments, thus optimizing natural ventilation design [36]. Orthogonal experimental design reduces the number of experiments and associated costs by selecting representative experimental points, while maintaining the universality and representativeness of the results. This approach simplifies the study of complex problems, allowing for the extraction of the influence of multiple factors at a limited number of points. The experimental results are easily analyzed, and orthogonal design facilitates efficient comparisons of the impacts of various factors using statistical methods such as range analysis and variance analysis, thereby helping to identify the optimal solution in multi-factor environments [37].

The combination of CFD and orthogonal experimental design provides a more comprehensive and efficient optimization strategy for the design of building natural ventilation. CFD delivers accurate airflow simulations, while orthogonal experimental design streamlines the experimental process and parameter selection. Together, they complement each other, enhancing both design efficiency and research accuracy [38].

A 3D steady-state CFD model was used in this study. The simulation adopted the standard k-ε turbulence model and neglected temperature variation. From Table 4, we can see the computational domain was extended to $5 \times 5 \times 3$ times the building size. Boundary conditions included velocity inlets based on local meteorological wind data and pressure outlets on the downwind sides. Local mesh refinement was implemented near building surfaces to improve accuracy. A convergence criterion of 10^{-4} residuals was applied with a total of 1000 iterations.

Table 4. Simulation parameter setting.

Parameter	Specific Setting	Parameter	Specific Setting
Model creation	RHINO modeling into PHOENICS	Selection governing equations	Standard k-ε turbulence model
Computational domain	Length × width × height: 5X × 5Y × 3Z	Meteorological parameters	Wind direction: East–Northeast Wind speed: 1.9 m/s in summer, 1.4 m/s in winter
Number of iterations	1000	Grid division	Building local encryption

3. Results and Analysis

3.1. Outdoor Wind Environment Simulation Analysis

As shown in Table 2, the prevailing wind direction in Xi'an during both summer and winter originates from the northeast. Although seasonal variations in wind speed are observed, the overall characteristics of the outdoor wind environment remain consistent across seasons. Therefore, for the purpose of analysis, this study takes the summer condition as a representative scenario. The simulation results at a pedestrian height of 1.5 m are analyzed for the selected case building, as illustrated in Figure 1.

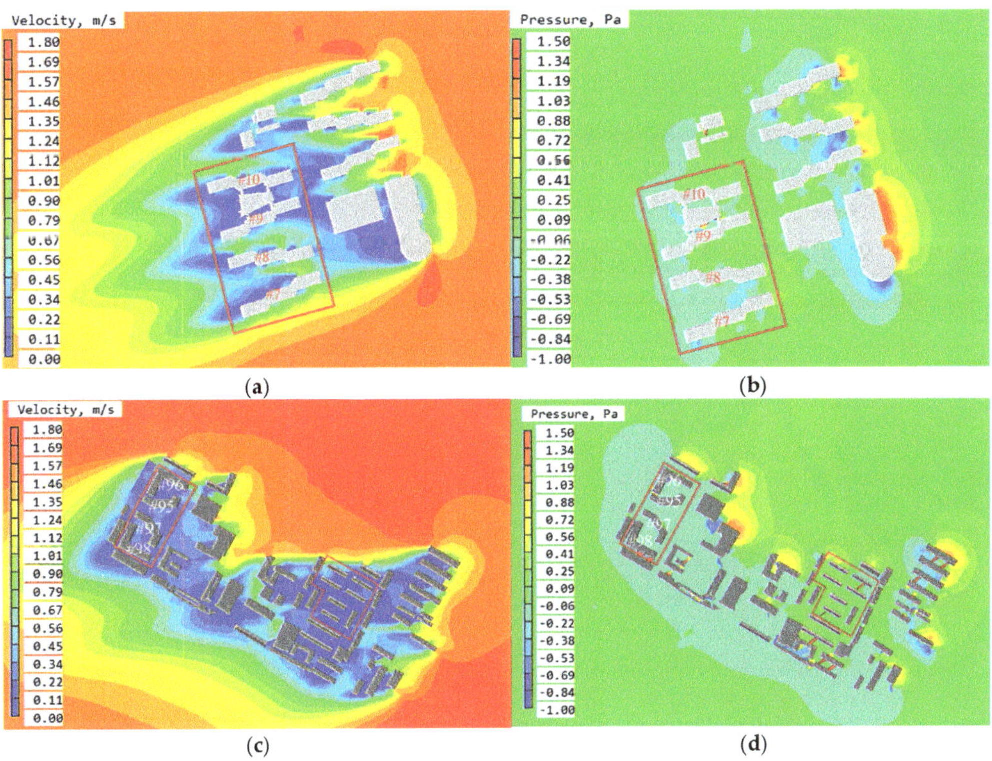

Figure 1. Wind Speed and Wind Pressure Contour Maps at Pedestrian Height of Plate-Type Multi-Story Building. (**a**) Plate-type multi-story wind speed cloud map of the new campus of CHD. (**b**) Plate-type multi-story wind pressure cloud map of the new campus of CHD. (**c**) Plate-type multi-story wind speed cloud map of XD new campus. (**d**) Plate-type multi-story wind pressure cloud chart of XD new campus.

Based on the wind speed contours (a) and (c) and the wind pressure contours (b) and (d), the common issues in the east and west districts of the XD new campus and the east district of the CHD new campus can be summarized as follows:

1. Wind Speed Issue

In all three areas, the dormitory areas generally exhibit low wind speeds, particularly in regions with enclosed or staggered layouts, where airflow is obstructed, resulting in large calm wind zones. For example, on the leeward face of Building No. 95 in the western district of XD, a wind shadow is formed. Similarly, the enclosed courtyards of Buildings No. 97 and No. 98 experience poor ventilation. A similar reduction in wind speeds is observed in both the eastern district of XD and the eastern district of CHD. Notably, in areas exhibiting the narrow tube effect between dormitory buildings, local wind speeds are slightly enhanced. However, the overall impact remains limited, failing to significantly improve the overall ventilation.

2. Wind Pressure Issue

The wind pressure contours show that the wind pressure on the windward face of the buildings is generally high, particularly at the building corners, where there is a significant wind pressure differential. For instance, the wind pressure difference before and after the first row of buildings in the western district of XD is 1.35 Pa, while the maximum wind pressure difference in the eastern district reaches 1.56 Pa. Similarly, the wind pressure difference in the eastern district of CHD also reaches 1.56 Pa. Although the wind pressure inside the dormitory areas remains relatively stable, the pressure difference before and after the buildings is generally small, leading to poor airflow and limited ventilation. Particularly in low wind speed conditions, the wind pressure and wind speed are proportional, which hampers the effective enhancement of natural ventilation within the buildings.

3. Poor Ventilation

Whether in the western or eastern districts of XD, or in the eastern district of CHD, the dormitory areas generally suffer from poor ventilation. This issue is particularly pronounced in the summer months, when internal wind pressure is relatively low, resulting in stuffy conditions that adversely affect ventilation. The enclosed layouts and internal blockages within the buildings hinder the free flow of air, leading to areas with insufficient ventilation [39].

In summary, both of these areas experience issues related to poor ventilation and uneven distribution of wind speed and pressure, which can primarily be attributed to the building layout and dimensions. In particular, airflow obstructions and wind pressure imbalances, resulting from the buildings' layout and dimensions, significantly impair the natural ventilation performance within the dormitory areas [40].

3.2. Measured Data Analysis and Simulation Verification

The multi-story buildings in the eastern area of the new campus of CHD and the multi-story buildings on the new campus of XD were measured on 10 January 2024, and 28 June 2024, respectively. Measurement points were placed at key locations, including the windward and leeward sides, the spaces between adjacent buildings, and at the building corners, to ensure comprehensive data collection. The test instruments used were the TES-1341 hot-wire anemometer (accuracy: ±3% of reading ±0.1 m/s) and the Testo 410-1 impeller anemometer (accuracy: ±1.5% of reading ±0.2 m/s), as shown in Figure 2. Wind speed measurements were conducted from 8:00 a.m. to 7:00 p.m., with data recorded every 0.5 h. The measured results for the eastern area of the new campus of CHD are presented in Figure 3.

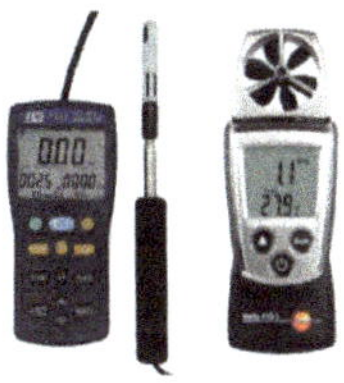

Figure 2. Test instruments: hot-wire anemometer and impeller anemometer.

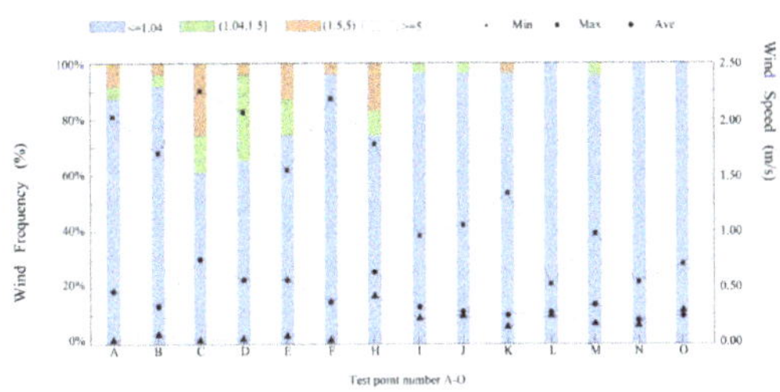

(a) Summer

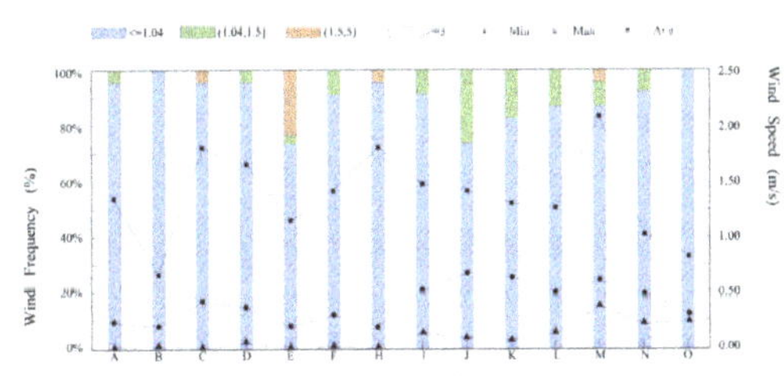

(b) Winter

Figure 3. Average wind speed, maximum and minimum wind speed, and wind frequency at all points in winter and summer on the new campus of CHD.

In the summer, the outdoor average wind speed throughout the day in the multi-story dormitory area is less than 1 m/s. Overall poor ventilation hinders pollutant diffusion, which adversely affects the health of students. In the winter, the maximum wind speed at measurement point M of the multi-story building in the eastern area of the new campus of CHD reaches 2.09 m/s. Based on the "Evaluation Standards for Green Campuses (GB/T 51356-2019 [41])", the wind speed in recreational and activity zones outdoor should not exceed 2 m/s. The wind speed at this measurement point exceeds the recommended limit, which is unfavorable for wind proofing and thermal insulation during winter [42]. Similarly, in the dormitory zone of the new campus of XD, poor ventilation is observed in the summer, and both wind proofing and heat preservation in the winter are similarly compromised.

This study employs two indicators, the Mean Absolute Percentage Error (MAPE) and the Root Mean Square Error (RMSE), to verify the accuracy of the simulation results. These are calculated using Formulas (1) and (2), respectively. In the formulas, Xsim represents the simulated value, Xmed represents the measured value, and N is the number of measurements. According to Figure 4, the average absolute percentage error for the slab-type multi-story dormitories is 15.54%, and the Root Mean Square Error is 8.22%, both of which fall within a reasonable error range. The results for both the eastern and western areas of the new campus of XD exhibit similar consistency, indicating that the model and simulation method accurately reflect the real conditions. Furthermore, by comparing the measured and simulated wind speed data graphically, it is evident that the overall trends align closely, with very high similarity at individual measurement points. The observed errors are attributed to factors such as the presence of pedestrians and vegetation during the measurements.

$$\text{MAPE} = \frac{\sum_{i=1}^{n} \frac{|X_{med} - X_{sim}|}{X_{med}}}{N}, \tag{1}$$

$$\text{RMSE} = \sqrt{\frac{\sum_{i=1}^{n} (X_{med} - X_{sim})^2}{N}}. \tag{2}$$

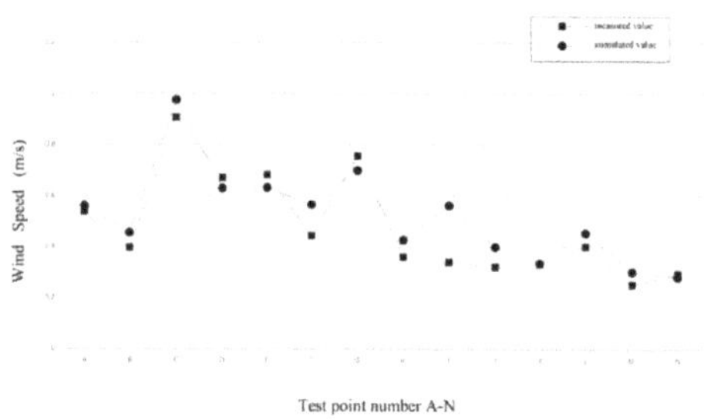

Figure 4. Comparison of measured and simulated wind speeds of CHD.

3.3. Orthogonal Experimental Design

The building layout of each scheme needs to set constraint conditions, mainly including fire prevention, sunshine and relevant planning requirements, as shown in Table 5. The simulated wind direction is northeast by east (ENE), which is the dominant wind direction in Xi'an [43]. The wind speed is a typical speed of 1.9 m/s.

Table 5. Scheme constraint conditions.

Constraint Name	Constraint Conditions	Basis
Dormitory design requirements	The spacing between dormitory buildings should meet the fire prevention, sunlight, and other relevant requirements, as well as comply with local urban planning regulations.	Code for design of dormitory building (JGJ36-2016) [44]
Fire prevention requirements	The minimum distance between multi-story civil buildings should be 9 m.	Code for fire protection design of buildings (GB-50016-2014) [45]
Sunshine requirements	Ensure that shaded buildings meet the sunlight requirement of ≥2 h on the winter solstice. The calculation starts from the bottom window sill, which is set at 0.9 m.	Building sunshine standard
Spacing requirements for gable walls	The minimum spacing between multi-story civil buildings on the gable side is 9 m.	Provisions on Urban Planning Administration of Xi'an City

The wind environment is influenced by various factors, including layout form, building size, geomorphic conditions, and wind direction. Among these, layout form and building size are particularly critical, as they are more easily optimized during the planning and design phases [46]. Therefore, for multi-story dormitory buildings, this study selects four key influencing factors, each set at three levels, as outlined in Table 6.

Table 6. Influencing factors and levels of outdoor wind environment of multi-story buildings.

	A. Building Layout	B. Building Length (m)	C. Building Width (m)	D. Building Height (m)
1	A_1 (Aligned layout)	B_1 (40)	C_1 (14)	D_1 (11.4)
2	A_2 (Staggered layout)	B_2 (50)	C_2 (16)	D_2 (18.6)
3	A_3 (Enclosed layout)	B_3 (60)	C_3 (18)	D_3 (22.2)

This study employs the L9 (4 × 3) orthogonal table for test design, as shown in Table 7. The orthogonal experiment matrix was generated and analyzed using SPSS Statistics 27, a widely used statistical software. SPSS was chosen for its robust data handling capabilities, flexibility in experimental design, and ease of statistical analysis [47].

Table 7. Orthogonal test table for plate-type multi-layer.

Simulation Schemes	Factor A: Building Layout	Factor B: Building Length (m)	Factor C: Building Width (m)	Factor D: Building Height (m)
1. $A_1B_1C_1D_1$	A_1 (Aligned layout)	B_1 (40)	C_1 (14)	D_1 (11.4)
2. $A_1B_2C_3D_2$	A_1	B_2 (50)	C_2 (16)	D_2 (18.6)
3. $A_1B_3C_2D_3$	A_1	B_3 (60)	C_3 (18)	D_3 (22.2)
4. $A_2B_1C_3D_3$	A_2 (Staggered layout)	B_1	C_3	D_3
5. $A_2B_2C_2D_1$	A_2	B_2	C_2	D_1
6. $A_2B_3C_1D_2$	A_2	B_3	C_1	D_2
7. $A_3B_1C_2D_2$	A_3 (Enclosed layout)	B_1	C_2	D_2
8. $A_3B_2C_1D_3$	A_3	B_2	C_1	D_3
9. $A_3B_3C_3D_1$	A_3	B_3	C_3	D_1

Figure 5 presents the CFD-simulated wind flow patterns under nine orthogonal design schemes, including determinant, staggered, and enclosed layouts with varying building dimensions. The comparative visualization of airflow distribution allows for intuitive evaluation of how different morphological parameters affect the near-ground wind environment. This figure serves as a key reference for supporting the analysis of wind environment characteristics across the tested configurations.

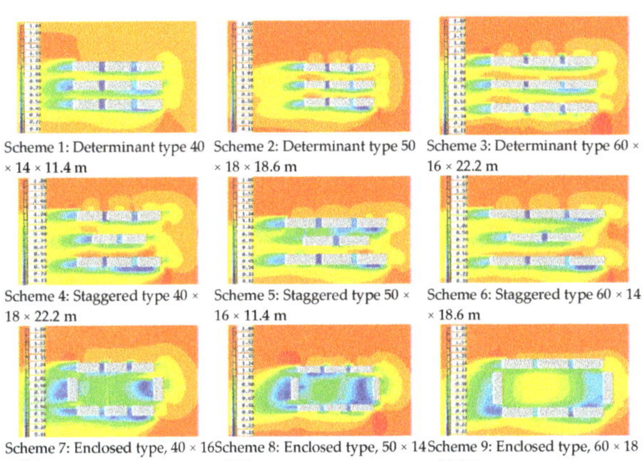

Figure 5. Simulation cloud diagrams of each orthogonal test scheme.

A total of 90 measurement points are selected from nine different schemes. These points are uniformly distributed across the windward face, leeward face, and between the gable walls of adjacent structures. The specific layout of the measurement points is shown in Figure 6.

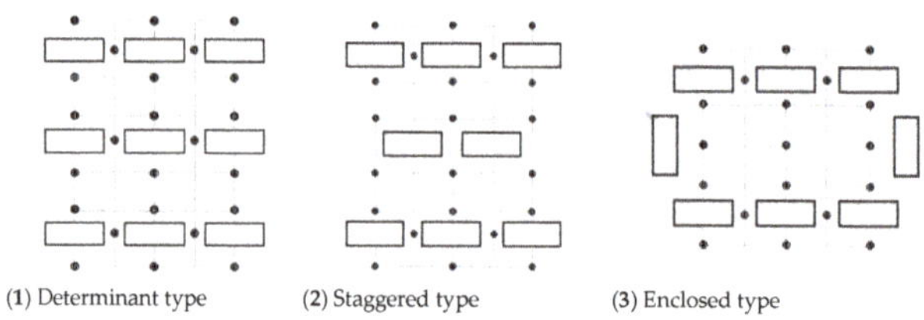

Figure 6. Schematic diagram of simulated measuring point layout.

As shown in Figure 7, the wind speed values at each simulated measurement point reflect the influence of building layout and size on wind speed. Generally, the wind speed on the windward face of buildings is relatively high, while the wind speed on the leeward face and between the gable walls of adjacent buildings is relatively low. In the determinant layout (Schemes 1–3), the wind speed distribution is uniform, and the wind speed on the windward face increases slightly as building width increases. In the staggered layout (Schemes 4–6), the overall wind speed is relatively high, especially on the leeward face and between the gable walls of adjacent buildings. As building height and length increase, wind speed is enhanced. The enclosed layout (Schemes 7–9) results in the lowest wind speed, which could result in inadequate ventilation. In general, building schemes with larger scale and width lead to higher wind speeds, while enclosed layouts tend to cause low wind speeds and insufficient ventilation.

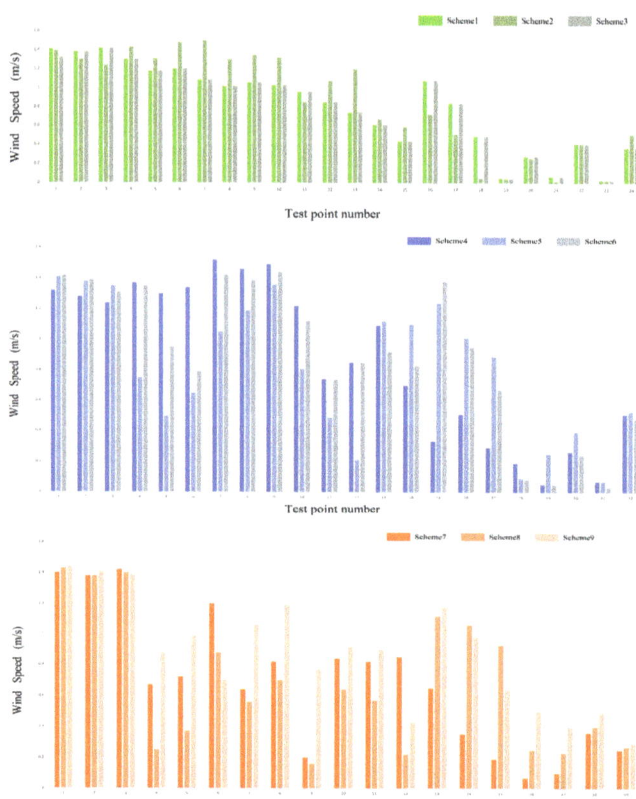

Figure 7. Wind speed values at simulated measurement points of multi-story dormitory.

By calculating the average wind speed at the simulated measurement points for each scheme, the simulation results of the orthogonal test for the multi-story dormitory block are presented in Table 8. It is observed that Scheme 3 yields the highest average wind speed, while Scheme 7 produces the lowest. Table 8 presents the simulation schemes established through an orthogonal experimental design conducted using SPSS software. The schemes investigate the influence of various building layouts—namely determinant, staggered, and enclosed configurations—on the average wind speed within the building cluster. The aim is to identify an optimal spatial arrangement that promotes favorable wind flow conditions.

Table 8. Table of simulation results of plate-type multilayer orthogonal experiment.

Simulation Schemes	A: Building Layout	B: Building Length (m)	C: Building Width (m)	D: Building Height (m)	(V) Average Wind Speed (m/s)
1. $A_1B_1C_1D_1$	A_1 (Aligned layout)	B_1 (40)	C_1 (14)	D_1 (11.4)	0.71
2. $A_1B_2C_3D_2$	A_1	B_2 (50)	C_2 (16)	D_2 (18.6)	0.76
3. $A_1B_3C_2D_3$	A_1	B_3 (60)	C_3 (18)	D_3 (22.2)	1.03
4. $A_2B_1C_3D_3$	A_2 (Staggered layout)	B_1	C_3	D_3	0.81
5. $A_2B_2C_2D_1$	A_2	B_2	C_2	D_1	0.98
6. $A_2B_3C_1D_2$	A_2	B_3	C_1	D_2	1.02
7. $A_3B_1C_2D_2$	A_3 (Enclosed layout)	B_1	C_2	D_2	0.56
8. $A_3B_2C_1D_3$	A_3	B_2	C_1	D_3	0.76
9. $A_3B_3C_3D_1$	A_3	B_3	C_3	D_1	0.93

Based on the average wind speed values, the statistical parameters K, Kavg, and R are calculated, as presented in Table 9. According to the magnitude of R, the factors influencing wind speed are ranked as follows: building width > building length > layout form > building height. Therefore, in the planning and design of multi-story buildings, building width should be the primary consideration, followed by building length and layout form, with building height having relatively little impact on the outdoor wind environment.

Table 9. Range table of plate-type multilayer orthogonal experiment.

	Levels	A. Building Layout	B. Building Length (m)	C. Building Width (m)	D. Building Height (m)
K	1	2.5	2.08	2.49	2.62
	2	2.81	2.5	2.57	2.34
	3	2.25	2.98	2.5	2.6
K_{avg}	1	0.833	0.693	0.415	0.873
	2	0.934	0.833	0.857	0.780
	3	0.750	0.993	0.833	0.867
Optimal Level		A_2	B_2	C_2	D_1
R		0.184	0.300	0.442	0.093
Optimization percentage		A_3 (Enclosed layout)	B_3 (60)	C_3 (18)	D_3 (22.2)

The comfortable wind speed range of 1–2 m/s is used as the evaluation criterion. the optimal combination for the multi-story dormitory buildings is $A_2B_3C_2D_1$, corresponding to a staggered layout, a building length of 60 m, a width of 16 m, and a height of 11.4 m.

3.4. Influencing Factor Analysis

The wind speed characteristics in the built environment of cold regions should comply with the provisions outlined in the "Assessment Standard for Green Buildings (GB/T 50378-2019)" [48]. In the section on environmental livability, the pedestrian activity areas on the site should not contain vortex or windless zones under typical summer wind conditions. During typical winter wind conditions, the wind speed at a height of 1.5 m in the pedestrian area around the building should be less than 5 m/s [49], while the wind speed in outdoor rest areas should be less than 2 m/s [20]. Additionally, according to relevant studies, people tend to feel comfortable when the wind speed exceeds 1 m/s. Therefore, the comfortable wind speed range is set between 1 and 2 m/s, and the coupling relationship analysis is conducted based on this standard [50].

According to the results of the orthogonal experiments, the relationship between the influencing factors and the wind environment is shown in Figure 8. For multi-story dormitory areas, the wind environment index is highest when the building layout is staggered. The average wind speeds, in descending order, are as follows: staggered form > determinant form > enclosed form. The building length is directly proportional to wind speed, as the building length increases, so does the wind speed. When the building width increases from 14 m to 18 m, the wind speed initially increases, then decreases. Similarly, an increase in building height from three to six stories leads to an initial decrease followed by a subsequent increase in wind speed.

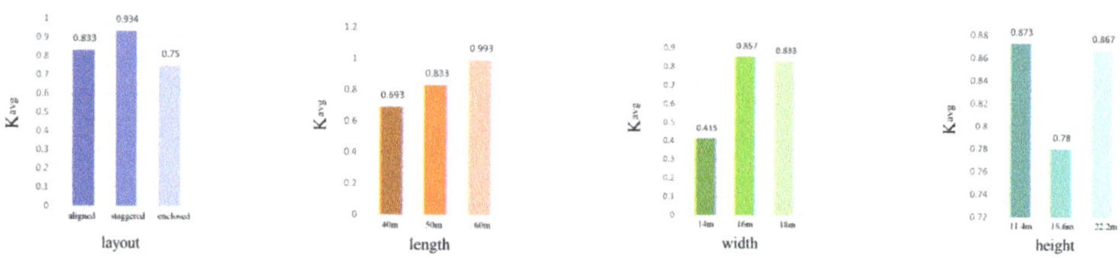

Figure 8. Analysis of the advantages and disadvantages of different levels of various influencing factors.

3.5. Optimal Scheme Verification

In this study, based on the optimal combination scheme for multi-story buildings, a layout optimization simulation is conducted for the dormitory area in the eastern district of the new CHD campus. The optimization scheme adopts the optimal combination of multi-story buildings as $A_2B_3C_2D_1$, which corresponds to a staggered layout, a building length of 60 m, a width of 16 m, and a height of 11.4 m. By simulating and analyzing the airflow within the dormitory area under the optimal layout, as shown in Figures 9 and 10, the following conclusions are drawn:

1. Uniform Wind Speed Distribution.

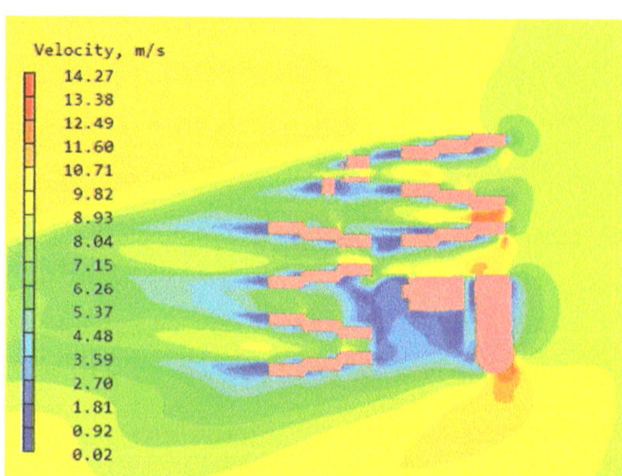

Figure 9. Schematic diagram of outdoor wind environment simulation of original building layout.

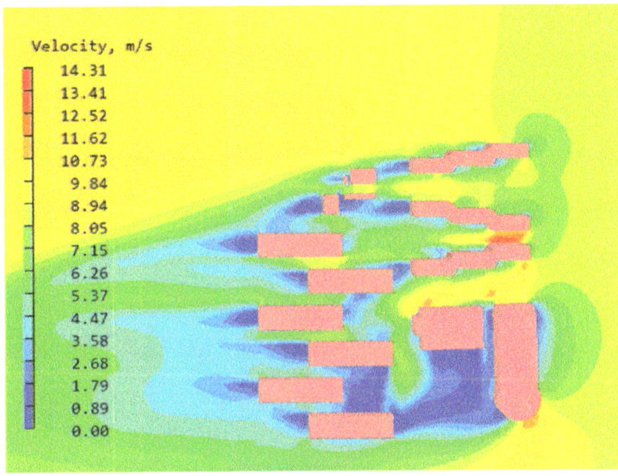

Figure 10. Schematic diagram of outdoor wind environment simulation of adjusted building layout.

The optimized design results in a relatively uniform wind speed distribution without extreme peaks. The airflow is smooth, and the fluctuation in wind speed is minimal, thus preventing discomfort caused by strong winds.

2. Improvement in Human Comfort.

Through layout and size optimization, areas of high wind speed are minimized, reducing the potential for wind-related discomfort and enhancing overall comfort within the dormitory area.

3. Good Ventilation Effect.

The spacing and arrangement of the buildings are suitable, with no significant wind speed dead zones or areas with excessively high wind speeds. This ensures effective ventilation and smooth airflow within the building complex, promoting good air mobility.

3.6. Strategy Analysis Summary

1. Staggered Layout Design.

The staggered layout is a core design strategy in this study, aimed at reducing direct convective obstruction between buildings through the staggered arrangement [51]. This layout prevents buildings from facing the wind directly by disrupting potential wind speed dead zones that may form between them. As a result, it not only mitigates the risk of excessively high or stagnant wind speeds but also significantly enhances airflow smoothness [52]. Consequently, the wind speed distribution within the dormitory area becomes more uniform, and airflow is more continuous, leading to a substantial improvement in ventilation efficiency.

2. Optimization of Building Dimensions.

In the optimal scheme, a building dimension combination of 60 m in length, 16 m in width, and 11.4 m in height is selected. These dimensions are designed to avoid extreme wind conditions, preventing both excessively high and low wind speeds. The optimized building dimensions effectively alleviate discomfort caused by excessive wind speeds, while maintaining adequate airflow to support a comfortable living environment. This design strategy not only reduces the extent of high-wind-speed zones but also lowers the risk of wind-induced hazards to residents, thereby enhancing overall living comfort.

3. Avoidance of Wind Speed Dead Zones and High Wind Speed Obstacles.

In the optimal layout, no obvious wind speed dead zones are formed between building gaps and areas. Wind speed dead zones typically result from an unreasonable building layout or insufficient spacing between buildings, leading to airflow stagnation and reduced ventilation efficiency [5]. By adopting a staggered layout and rational building dimensions, the design enhances inter-building airflow. It effectively avoids airflow disruptions due to extreme wind speeds and substantially improves the ventilation performance of the dormitory area.

4. Improved Ventilation Effect.

By incorporating a staggered layout and optimizing building dimensions, the dormitory area achieves a more balanced wind speed distribution and smoother airflow. The wind speed fluctuations are relatively mild, preventing extreme variations. This leads to improved ventilation efficiency and reduced discomfort caused by excessively high or low wind speeds. Overall, the optimized design ensures optimal ventilation within the multi-story dormitory layout, fostering a more comfortable living environment.

4. Discussion

4.1. Comparison with the Existing Literature

The preceding literature review categorized representative studies based on building layout forms, dimensions, and shapes, as summarized in Table 1. However, a comparison with Table 10—which lists studies specifically focusing on multi-story dormitory buildings—reveals that such studies are relatively few in number. This contrast underscores that, despite some recent efforts, the majority of existing research remains concentrated on residential and block-type buildings, with limited attention paid to dormitory-specific wind environment analysis, particularly in cold regions.

Table 10. Representative studies on the optimization of outdoor wind environment in multi-story dormitories.

Research Factors	Building Types	Specific Factors	Literature	Research Methods
Layout form	Campus architecture	Influence of layout mode on wind environment	[6–9,53]	Numerical simulation
Building dimensions	Campus architecture	Influence of building height and width on wind environment	[54]	Wind tunnel test
Building forms	Campus architecture	Influence of building shape on wind environment	[55]	Numerical simulation
		Influence of building shape on wind environment	[50]	Numerical simulation + On-site measurement

Tie et al. examined winter wind conditions in dormitory areas at Hohhot University using CFD simulations. Their results show that diagonal and enclosed layouts of U- and L-shaped buildings improve outdoor comfort, offering climate-adaptive design strategies for cold region dormitories [6]. Zhang et al. investigated pedestrian wind comfort around lift-up buildings with varying aspect ratios and core configurations. Wind tunnel tests revealed that recessed core designs reduce high-wind zones, providing guidance for urban morphological optimization [54]. Janssen et al. compared multiple pedestrian wind comfort criteria using CFD simulations around university buildings. Their study revealed inconsistencies among prevailing standards, emphasizing the need for more unified evaluation frameworks [50].

These studies highlight the crucial role of building form and layout in shaping outdoor wind environments, while also revealing the scarcity of research focused on dormitory settings in cold regions.

4.2. Limitations and Future Research

To address this gap, the present study investigates the outdoor wind environment of multi-story dormitory buildings in a cold-region context using an integrated approach that combines field measurements with CFD simulations. This methodology enhances the accuracy of the analysis and provides a practical framework for assessing wind performance in similar climatic conditions.

Nevertheless, this study has the following limitations:

(1) The simulation only covers winter and summer conditions in Xi'an, without considering transitional seasons. Future studies should incorporate full-year climatic data to support more comprehensive and accurate evaluations of outdoor wind comfort.

(2) The analysis focuses on layout form, building dimensions, and shape, but does not account for other potential factors such as building density and morphological complexity. These variables should be integrated into future models for a more robust prediction of wind performance.

(3) The study concentrates on the outdoor wind environment of dormitory buildings and does not address other aspects of the physical environment, such as indoor ventilation or thermal comfort. A more holistic approach is needed to guide integrated improvements in campus living environments.

5. Conclusions

This study investigates the optimization of the outdoor wind environment in multi-story university dormitory areas in cold regions, using CFD simulation combined with orthogonal experimental design. The work is grounded in architectural practice, aiming to inform early-stage design decisions rather than relying solely on theoretical modeling.

(1) Building layout and dimensional parameters (length, width, height) significantly influence pedestrian-level wind conditions, with building width identified as the most critical factor.

(2) The optimal configuration—a staggered layout with dormitory dimensions of 60 m (length), 16 m (width), and 11.4 m (height)—enhances airflow uniformity and spatial balance.

(3) CFD results from the CHD show that this scheme reduces near-ground wind speed by 18%, improving outdoor comfort in both summer and winter.

(4) The findings underscore the importance of considering not only individual parameters, but also the interaction between building form and wind flow in design evaluation.

By aligning environmental simulation with architectural thinking, this study provides a practical reference for climate-responsive dormitory design and supports the development of sustainable campus environments in cold regions.

Author Contributions: Conceptualization, X.C. and Q.L.; methodology, X.C., H.K. and Q.L.; software, X.C.; validation, X.C., H.K. and J.Z.; formal analysis, X.C.; investigation, H.K. and J.Z.; resources, X.C. and H.K.; data curation, X.C. and H.K.; writing—original draft preparation, X.C. and Q.L.; writing—review and editing, Q.L.; visualization, H.K. and J.Z.; supervision, X.C. and Q.L.; project administration, Q.L.; funding acquisition, Q.L. All authors have read and agreed to the published version of the manuscript.

Funding: This research was funded by Shaanxi Provincial Department of Science and Technology, the key international cooperation projects (grant number 2024GH-ZDXM-03).

Data Availability Statement: The data presented in this study are available on request from the corresponding author.

Conflicts of Interest: The authors declare no conflicts of interest.

References

1. Du, Y.X.; Mak, C.M. Improving Pedestrian Level Low Wind Velocity Environment in High-Density Cities. *Sustain. Cities Soc.* **2018**, *42*, 314–324. [CrossRef] [PubMed]
2. Alnusairat, S.; Al-Shatnawi, Z.; Ayyad, Y.; Alwaked, A.; Abuanzeh, N. Rethinking Outdoor Courtyard Spaces on University Campuses to Enhance Health and Wellbeing: The Anti-Virus Built Environment. *Sustainability* **2022**, *14*, 5602. [CrossRef]
3. Liu, Q.; Ren, J. Research on the building energy efficiency design strategy of Chinese universities based on green performance analysis. *Energy Build.* **2020**, *224*, 110242. [CrossRef]
4. Shui, T.T.; Liu, J.; Yuan, Q.; Qu, Y.; Jin, H.; Cao, J.L.; Liu, L.; Chen, X. Assessment of pedestrian-level wind conditions in severe cold regions of China. *Build. Environ.* **2018**, *135*, 53–67. [CrossRef]
5. Myroniuk, K.; Furdas, Y.; Zhelykh, V.; Yurkevych, Y. Examining Wind Flow's Impact on Multi-Storey Buildings: A Quest for Quality Improvement. *Prod. Eng. Arch.* **2024**, *30*, 57–66. [CrossRef]
6. Tie, W.; Wang, W.; Lin, D. Enhancing the winter wind conditions in the dormitory vicinity of Hohhot University through a biomechanical perspective. *Mol. Cell. Biomech.* **2025**, *22*, 1076. [CrossRef]
7. Deng, J.Y.; Xia, Y.; Lao, H.; Ye, Y.K.; Wang, Z.J.; Jiang, H.Y. Natural ventilation potential of teaching building complexes with different block shapes and layout patterns. *J. Build. Eng.* **2024**, *96*, 110420. [CrossRef]
8. Mei, S.J.; Hu, J.T.; Liu, D.; Zhao, F.Y.; Li, Y.G.; Wang, Y.; Wang, H.Q. Wind driven natural ventilation in the idealized building block arrays with multiple urban morphologies and unique package building density. *Energy Build.* **2017**, *155*, 324–338. [CrossRef]
9. Wu, Z.G.; Li, B.L.; Shi, F.; Xiao, Z.H.; Hong, X.C. Analysis of the Impact of Layout Mode on the Wind Environment of Dormitories in Coastal Universities in Southern Fujian Province. *Buildings* **2023**, *13*, 3030. [CrossRef]
10. Poddaeva, O.; Churin, P. Improving the comfort of pedestrian and recreational areas of the stylobate part of the residential complex. *Energy Rep.* **2021**, *7*, 395–402. [CrossRef]
11. Zheng, S.; Wang, Y.; Zhai, Z.Q.; Xue, Y.; Lin, D.M. Characteristics of wind flow around a target building with different surrounding building layers predicted by CFD simulation. *Build. Environ.* **2021**, *201*, 107962. [CrossRef]
12. Ma, T.; Chen, T. Classification and pedestrian-level wind environment assessment among Tianjin's residential area based on numerical simulation. *Urban Clim.* **2020**, *34*, 100702. [CrossRef]
13. Mou, B.; He, B.J.; Zhao, D.X.; Chau, K.W. Numerical simulation of the effects of building dimensional variation on wind pressure distribution. *Eng. Appl. Comput. Fluid Mech.* **2017**, *119*, 293–309. [CrossRef]
14. Zahid Iqbal, Q.M.; Chan, A.L.S. Pedestrian level wind environment assessment around group of high-rise cross-shaped buildings: Effect of building shape, separation and orientation. *Build. Environ.* **2016**, *101*, 45–63. [CrossRef]
15. Wu, Y.H.; Li, W.F.; Zeng, N.Y.; Bai, X.X. Analyzing the Impact of Tall Building Geometries on Wind Environment in a Hypothetical Urban Context: A Typological and Parametric Study. *Lect. Notes Civ. Eng.* **2023**, *559*, 565–577.
16. Dai, Y.W.; Mak, C.M.; Ai, Z.T. Computational fluid dynamics simulation of wind-driven inter-unit dispersion around multi-storey buildings: Upstream building effect. *Indoor Built Environ.* **2017**, *28*, 217–234. [CrossRef]
17. Druenen, V.T.; Hooff, T.V.; Montazeri, H.; Blocken, B. CFD evaluation of building geometry modifications to reduce pedestrian level wind speed. *Build. Environ.* **2019**, *163*, 106293. [CrossRef]
18. Xie, Z.Y.; Yang, N. Optimization Design Strategy of High-Rise Building Form for Improving Outdoor Wind Environment. *Archit. J.* **2013**, 2.
19. Adamek, K.; Vasan, N.; Elshaer, A.; English, E.; Bitsuamlak, G. Pedestrian level wind assessment through city development: A study of the financial district in Toronto. *Sustain. Cities Soc.* **2017**, *35*, 179–190. [CrossRef]
20. Hong, B.; Lin, B.R. Numerical studies of the outdoor wind environment and thermal comfort at pedestrian level in housing blocks with different building layout patterns and trees arrangement. *Renew. Energy* **2015**, *73*, 18–27. [CrossRef]
21. Shirzadi, M.; Tominaga, Y. CFD evaluation of mean and turbulent wind characteristics around a high-rise building affected by its surroundings University campus. *Build. Environ.* **2022**, *225*, 109637. [CrossRef]
22. Le, D.; Li, N.P.; Su, L. Study on wind environment characteristic of different height high-rise residential buildings in Changsha, China. *Nat. Environ. Pollut. Technol.* **2016**.
23. Taseiko, O.V.; Mikhailuta, O.V.; Pitt, A.; Lezhenin, A.A.; Zakharov, Y.V. Air pollution dispersion within urban street canyons. *Atmos. Environ.* **2009**, *43*, 245–252. [CrossRef]
24. Poddaeva, O.; Churin, P. Numerical simulation of the pedestrian comfort of the micro district. *Sci. Direct.* **2022**, *8*, 1491–1500.
25. Shi, X.; Zhu, Y.Y.; Duan, J.; Shao, R.Q.; Wang, J.G. Assessment of pedestrian wind environment in urban planning design. *Landsc. Urban Plan.* **2015**, *140*, 17–28. [CrossRef]

26. Kaseb, Z.; Hafezi, M.; Tahbaz, M.; Delfani, D. A framework for pedestrian-level wind conditions improvement in urban areas: CFD simulation and optimization. *Build. Environ.* **2020**, *184*, 107191. [CrossRef]
27. Tsichritzis, L.; Nikolopoulou, M. The effect of building height and façade area ratio on pedestrian wind comfort of London. *J. Wind Eng. Ind. Aerodyn.* **2019**, *191*, 63–75. [CrossRef]
28. Tse, K.T.; Weerasuriya, A.U.; Zhang, X.L.; Li, S.W.; Kwok, K.C.S. Effects of twisted wind flows on wind conditions in passages between buildings. *J. Wind Eng. Ind. Aerodyn.* **2017**, *167*, 87–100. [CrossRef]
29. Ramponi, R.; Blocken, B.; Laura Bde, C.; Janssen, W.D. CFD simulation of outdoor ventilation of generic urban configurations with different urban densities and equal and unequal street widths. *Build. Environ.* **2015**, *92*, 152–166. [CrossRef]
30. Chen, L.; Hang, J.; Sandberg, M.T.; Claesson, L.; Sabatino, S.D.; Wigo, H. The impacts of building height variations and building packing densities on flow adjustment and city breathability in idealized urban models. *Build. Environ.* **2017**, *118*, 334–361. [CrossRef]
31. *GB 50352-2019*; Unified Design Standards for Civil Buildings. Architecture & Building Press: Beijing, China, 2019.
32. CBA Clima Tool. Available online: https://clima.cbe.berkeley.edu/ (accessed on 10 January 2025).
33. *GB/T 19201-2006*; Grade of Tropical Cyclones. China National Standardization Management Committee: Beijing, China, 2006.
34. Lei, Z.; Liu, C.; Wang, L.; Li, N. Effect of natural ventilation on indoor air quality and thermal comfort in dormitory during winter. *Build. Environ.* **2017**, *125*, 240–247. [CrossRef]
35. World Meteorological Organization. Available online: https://public.wmo.int/en (accessed on 10 January 2025).
36. Jiru, T.E.; Bitsuamlak, G.T. Application of CFD in Modelling Wind-Induced Natural Ventilation of Buildings—A Review. *Int. J. Vent.* **2010**, *9*, 131–147. [CrossRef]
37. Zhu, J.J.; Chew, D.A.; Lv, S.; Wu, W.W. Optimization method for building envelope design to minimize carbon emissions of building operational energy consumption using orthogonal experimental design (OED). *Habitat Int.* **2013**, *37*, 148–154. [CrossRef]
38. Lu, Y.; Wang, T.; Zhao, C.; Zhu, Y.; Jia, X.; Zhang, L.; Shi, F.; Jiang, C. An efficient design method of indoor ventilation parameters for high-speed trains using improved proper orthogonal decomposition reconstruction. *J. Build. Eng.* **2023**, *71*, 106600. [CrossRef]
39. Ren, C.; Wang, J.; Feng, Z.; Kim, M.K.; Haghighat, F.; Cao, S.-J. Refined design of ventilation systems to mitigate infection risk in hospital wards: Perspective from ventilation openings setting. *Environ. Pollut.* **2023**, *333*, 122025. [CrossRef]
40. Yang, T.; Clements-Croome, D.J. Natural Ventilation in Built Environment. *Sustain. Built Environ.* **2018**, *23*, 434–464.
41. *GB/T 51356-2019*; Assessment Standard for Green Campus. Ministry of Housing and Urban-Rural Development of the People's Republic of China: Beijing, China, 2019.
42. Abdollahzadeh, N.; Biloria, N. Outdoor thermal comfort: Analyzing the impact of urban configurations on the thermal performance of street canyons in the humid subtropical climate of Sydney. *Front. Archit. Res.* **2021**, *10*, 394–409. [CrossRef]
43. Weather Spark. Available online: https://zh.weatherspark.com/ (accessed on 10 January 2025).
44. *JGJ36-2016*; Code for Design of Dormitory Building. Ministry of Housing and Urban-Rural Development: Beijing, China, 2017.
45. *GB-50016-2014*; Fire Protection Code for Building Design. Ministry of Housing and Urban-Rural Development: Beijing, China, 2015.
46. Yang, S.; Zhou, D.; Wang, Y. Comparing impact of multi-factor planning layouts in residential areas on summer thermal comfort based on orthogonal design of experiment (ODOE). *Build. Environ.* **2020**, *182*, 107145. [CrossRef]
47. Rahman, A.; Muktadir, M.G. SPSS: An Imperative Quantitative Data Analysis Tool for Social Science Research. *Int. J. Res. Innov. Soc. Sci.* **2021**, *5*, 300–301. [CrossRef]
48. *GB/T 50378-2019*; Green Building Evaluation Standard. Ministry of Housing and Urban-Rural Development: Beijing, China, 2024.
49. Mittal, H.; Sharma, A.; Gairola, A. A review on the study of urban wind at the pedestrian level around buildings. *J. Build. Eng.* **2018**, *18*, 154–163. [CrossRef]
50. Janssen, W.D.; Blocken, B.; Hooff, T.V. Pedestrian wind comfort around buildings: Comparison of wind comfort criteria based on whole-flow field data for a complex case study. *Build. Environ.* **2013**, *59*, 547–562. [CrossRef]
51. You, W.; Shen, J.L.; Ding, W.W. Improving residential building arrangement design by assessing outdoor ventilation efficiency in different regional spaces. *Archit. Sci. Rev.* **2018**, *61*, 202–214. [CrossRef]
52. Cheung, J.O.P.; Liu, C.H. CFD simulations of natural ventilation behaviour in high-rise buildings in regular and staggered arrangements at various spacings. *Energy Build.* **2011**, *43*, 1149–1158. [CrossRef]
53. Fan, L.Q.; Ren, S.W.; Ma, Y.; Liu, Q.B. The Coupling Relationship between Building Morphology and Outdoor Wind Environment in the High-Rise Dormitory Area in China. *Energies* **2023**, *16*, 3655. [CrossRef]
54. Zhang, X.L.; Tse, K.T.; Weerasuriya, A.U.; Li, S.W.; Kwok, K.C.S.; Mak, C.M.; Niu, J.L.; Lin, Z. Evaluation of pedestrian wind comfort near 'lift-up' buildings with different aspect ratios and central core modifications. *Build. Environ.* **2017**, *124*, 245–247. [CrossRef]
55. Du, Y.X.; Mak, C.M.; Liu, J.L.; Xia, Q.; Niu, J.L.; Kwok, K.C.S. Effects of lift-up design on pedestrian level wind comfort in different building configurations under three wind directions. *Build. Environ.* **2017**, *117*, 84–99. [CrossRef]

buildings

MDPI

Review

Enhancing Outdoor Environmental Comfort: A Review of Façade-Surface Strategies and Microclimate Impacts

Zahida Khan [1,*,†] and Mehdi Ghiai [2,*,†]

1 College of Architecture and Planning, Ball State University, Muncie, IN 47306, USA
2 Department of Design, Texas Tech University, Lubbock, TX 79409, USA
* Correspondence: zahida.khan@bsu.edu (Z.K.); mehdi.ghiai@ttu.edu (M.G.)
† These authors contributed equally to this work.

Abstract: Building façades traditionally focus on enhancing indoor environmental quality and improving energy performance, but undermine their influence on Outdoor Environmental Comfort (OEC), including thermal, acoustic, and visual conditions. With technological advancements in envelope design, research on new materials and green systems has been introduced in the last few decades. This review examines the role of two key elements—façade materials and green façades—in shaping OEC. A total of 41 peer-reviewed studies (24 on urban scale and 17 on building scale) were categorized into three focus areas: (1) outdoor thermal comfort; (2) outdoor acoustic comfort; and (3) outdoor visual comfort. The analysis was structured across three levels: (a) Performance Determinants; (b) Metrics/Models; and (c) Material or Façade Types. We proposed this analytical structure to highlight the interactions between building façades and OEC domains (thermal, acoustic, and visual comfort). Our results showed façade treatment can impact all three comfort factors related to OEC, but trade-offs must be evaluated. Moreover, the findings highlighted that additional research is required to cover variations in both climate and context conditions, due to their close association with the OEC. Finally, the conceptual framework is presented to synthesize the three comfort domains for sustainable outdoor environments.

Keywords: outdoor comfort; building façade; materials; green façade; envelope

1. Introduction

Climate change and global warming are among the most pressing global challenges of the coming decades. [1]. Buildings significantly contribute to these issues through their energy consumption, emissions, and material choices. At the same time, global population growth is projected to reach 10.4 billion by 2100 [2], accompanied by intensified urbanization. As cities densify, the built environment increasingly alters Outdoor Environmental Comfort (OEC) as defined by microclimates, ambient noise, and visual glare [3]. These directly impact the outdoor comfort levels of people using outdoor urban spaces, which are key to the success of social life [4]. Understanding and improving outdoor comfort is therefore essential to achieving healthier, more sustainable urban communities.

OEC plays a vital role in shaping urban life, influencing how often and how comfortably people use outdoor spaces. It is determined by a combination of climatic conditions, urban morphology, and building elements—especially façades. Traditionally, façades have been designed to enhance indoor environmental quality and energy efficiency [5]. However, as the impacts of the Urban Heat Island (UHI) effect become more severe, its role in influencing outdoor comfort is gaining attention. While horizontal surfaces are often implicated in UHI, vertical façades also contribute meaningfully by altering heat retention, airflow, and solar exposure. Façade properties—such as reflectivity, thermal mass, and material texture—can significantly affect Outdoor Thermal Comfort (OTC) [6,7]. Similarly, façades influence Outdoor Acoustic Comfort (OAC) by amplifying or diffusing sound through their geometry and absorptive properties [8]. In terms of Outdoor Visual Comfort (OVC), material reflectance and surface color contribute to glare intensity and overall luminance conditions [9]. Taken together, the façade is not just a barrier but a key interface for shaping Outdoor Environmental Quality (OEQ).

In a façade assembly, the exterior surface treatment plays a key role in defining the OEQ. This treatment can be seen through the lens of façade materials and green façades [10]. Properties of façade materials, such as reflectivity, can impact outdoor OEC by contributing to UHI. Studies report that replacing conventional masonry with high-albedo or retroreflective finishes lowers façade-surface temperatures by roughly 10–14 °C and improves pedestrian thermal-stress indices by about 3 °C in hot-dry and humid-subtropical settings [11,12]. Similarly, studies on green façades showed that cooling through evapotranspiration and shading could drop the façade surface temperature by 2–4 °C, and acoustic absorption reduced the noise level by 40 dB [13–15]. Hence, both these categories encompass a variety of façade treatments that impact OEC.

Façade treatments span traditional masonry and advanced composites engineered for thermal, acoustic, and visual performance [5,6] In parallel, green façade strategies such as living walls and vertical greenery systems are increasingly used to improve outdoor comfort [16]. As we race against the continuous development of these design strategies in building envelope design, there is a need to review the advanced materials and vegetation strategies used in building façades. This review focuses on both material-based and vegetative façade interventions, highlighting their mechanisms and impacts on outdoor comfort.

Building façades are an important element of any urban fabric [17]. The material treatment of these façades can alter the OEQ near the immediate exterior zone of buildings to the public exterior realm of cities, such as urban canyons and plazas. Studies in topics of building façades and OEC have focused either on building scale or urban scale [9,18,19]. In either case, the impact is significant at a human comfort level and must be taken into consideration in the study. This study aims to identify the current state of the art in this topic.

Moreover, although numerous studies address individual aspects of outdoor comfort, few have synthesized the influence of façade design across multiple domains of OEC, namely OTC, OAC, and OVC. This review addresses that gap by providing an integrated analysis of how façade materials and green systems influence thermal, noise, and visual aspects of outdoor comfort levels. To support holistic decision-making in design, a conceptual framework is proposed to illustrate cross-domain interactions and trade-offs.

2. Background

To gain a comprehensive understanding of OEC, it is essential to understand each domain's technical definition and related metrics or indices. Below is a synopsis of each domain.

2.1. Outdoor Thermal Comfort

Thermal comfort refers to the human perception of ambient environmental conditions. According to ASHRAE, it is defined as "the condition of mind which expresses satisfaction with the surrounding thermal environment and is assessed by subjective evaluation" [20]. While it was initially studied in the context of indoor environments, thermal comfort in outdoor settings has gained increasing attention due to the compounded effects of climate change, urbanization, and the UHI phenomenon. Building façades play a significant role in influencing OTC by modifying microclimatic conditions and shaping the usability of outdoor spaces.

This section reviews twenty-three studies that examine the impact of façade materials on OTC. Similar to indoor thermal comfort research, OTC is influenced by climatic variables (air temperature, humidity, wind velocity, and mean radiant temperature), as well as physiological and behavioral factors. Over a hundred indices for assessing OTC have been proposed [21,22], of which four are widely used: Physiological Equivalent Temperature (PET), Universal Thermal Climate Index (UTCI), Predicted Mean Vote/Percentage People Dissatisfied (PMV/PPD), and Standard Effective Temperature (SET*)

1. Physiological Equivalent Temperature (PET) is based on the MEMI heat budget model using variables metabolic activity (M), physical work output (W), net radiation of the body (R), convective heat flow (C), latent heat used to evaporate water that diffuses as vapor through the skin (E_D), sum of heat for air heating and humidification (E_{Re}), heat loss via sweat evaporation (E_{Sw}), and body heat storage for heating or cooling (S) [23]. Its equation is written as:

$$M + W + R + C + E_D + E_{Re} + E_{Sw} + S = 0$$

2. Universal Thermal Climate Index (UTCI) is derived by coupling the thermo-physiological UTCI-Fiala multi-node model [24] and the clothing model [25] for a given meteorological condition. Key variables in its calculation include *air temperature (T_a), mean radiant temperature (T_r), water vapor pressure (p_a), wind speed (v_a), and deviation from the air temperature (Offset)*. The equation is written as below [26]:

$$\mathrm{UTCI}\ (T_a, T_r, v_a, p_a) = T_a + \mathrm{Offset}\ (T_a, T_r, v_a, p_a)$$

The heat-balance equation for the UTCI-Fiala model is based on the energy budget from: (1) overall body heat exchange through *metabolic rate activity (M), mechanical power (W)*, and *change in heat content of the body-storage (S)*; (2) peripheral (skin) heat exchanges through *turbulent flux of sensible heat (Q_H), radiation budget (Q^*)*, and *turbulent flux of latent (Q_L)*; (3) heat (passive diffusion of water vapor through the skin) through *turbulent flux of latent heat (sweat evaporation) (Q_{SW})*; (4) respiratory heat exchanges through *respiratory heat flux (sensible and latent) (Q_{Re})*; and (5) thermal environmental parameters, namely, air temperature (T_a), mean temperature (T_{mrt}), air speed relative to the body (v), and partial vapor pressure (e). Its equation is written as [27].

$$M - W - [Q_H\ (T_a, v) + Q^*\ (T_{mrt})] - [\ Q_L\ (e, v) + Q_{SW}\ (e, v)] - Q_{Re}\ (T_a, v) \pm S = 0$$

3. Predicted Mean Vote/Percentage People Dissatisfied (PMV/PPD) was derived by O. P. Fanger [28] using variables, namely *metabolic rate (M), external work (W), and total heat loss from the body (H_{loss})*. The PMV equation in its simple form is given as

$$\mathrm{PMV} = [0.303 \times e^{-0.036M} + 0.028] \times [(M - W) - H_{loss}]$$

$Heat_{loss}$ is computed from convective heat loss (Qc), radiative heat loss (Qr), evaporative heat loss from skin (Qe), and heat loss through respiration (Qres), and is written as

$$H_{loss} = Qc + Qr + Qe + Qres$$

4. Standard Effective Temperature (SET*) was derived in Berkeley by [29] and is computed using two-node thermoregulation models and heat-balance equations

2.2. Outdoor Acoustic Comfort

Acoustic comfort refers to the perceived well-being and satisfaction with the acoustical conditions of a given environment [30]. It is a critical factor in both indoor and outdoor settings, influencing how individuals experience and interact with their surroundings. In indoor environments, acoustic comfort pertains to a building's ability to minimize external noise and create a quiet, distraction-free space that supports productivity and social interaction [31]. Similarly, in outdoor environments, acoustic comfort plays a crucial role in shaping the urban experience, significantly affecting the public's enjoyment and overall perception of urban spaces [32]. A key element influencing acoustic comfort is the building envelope, which acts as a barrier against noise and contributes to creating acoustically balanced spaces. Numerous studies have highlighted the significant impact of building façades on occupants' acoustic experiences [33]. In urban settings, façades can enhance outdoor acoustic comfort by mitigating unwanted noise, and their effects can be classified into three main categories: sound reflection, sound absorption, and sound production [34]. Some of the common metrics used for the acoustic performance of building façades include *Sound Transmission Loss (STL), Sound-Pressure Level (SPL), Sound-Absorption Coefficient (α), and Noise-Reduction Coefficient (NRC)*. Calculations of each of these metrics are given below:

1. The Sound Transmission Loss (STL) could be calculated using the equation below:

$$\text{TL (dB)} = 10 \times \log (\text{W1}/\text{W2})$$

   where 'W1' is the sound power incident on a wall, and 'W2' is the sound power transmitted through a wall.

2. The Sound-Pressure Level (SPL) equation could be written as

$$\text{SPL (dB)} = 20 \times \log10 (\text{p}/\text{p0})$$

   where 'p' is the measured sound pressure and 'p0' is the reference sound pressure.

3. The Sound-Absorption Coefficient (α) is calculated as

$$\alpha = l_a / l_i$$

   where 'l_a' = the sound intensity absorbed by the material and 'l_i' = the incident sound intensity.

These metrics offer valuable insights into how various façade materials and configurations impact sound propagation and overall acoustic conditions. This section reviews the impact of façade materials and green walls on outdoor acoustic comfort. The findings offer a deeper understanding of how design interventions, such as material selection and vegetation integration, can mitigate noise and enhance urban soundscapes.

2.3. Outdoor Visual Comfort

Visual comfort refers to lighting conditions that prevent strain, discomfort, or distractions. The EN standard defines it as "the subjective state of visual health caused by the visual environment." [35]. Although horizontal illuminance remains the default design metric, field studies show that vertical eye illuminance and luminance-based indices, such as luminance ratios, Daylight Glare Probability (DGP), and Daylight Glare Index (DGI), correlate more strongly with perceived comfort. Preferred work-plane levels typically range from 100 to 3000 lux, with most occupants favoring levels between 300 and 2000 lux. Because no single index performs reliably across all daylight and electric-lighting conditions, DGI remains the most widely used practical benchmark, underscoring the need for a more transparent framework to guide metric selection and to develop more comprehensive visual-comfort indices [36,37].

While commonly studied indoors, visual comfort is equally important in outdoor environments. OVC results from the interaction between materials in the outdoor environment and lighting levels. Although both horizontal and vertical surfaces contribute to the visual comfort, this review focuses mainly on the latter—building façades. The focus of visual comfort studies has been either to mitigate the visual glare or to enhance visual pleasure. A comprehensive understanding of visual comfort in outdoor environments is best achieved through an examination of the state of the art in factors and performance indicators.

3. Methodology

This study is a comprehensive review that synthesizes findings from 41 peer-reviewed publications, including journal articles, conference proceedings, and review papers, to evaluate how building façades—particularly materials and vegetative systems—influence OEC. The review emphasized studies that addressed performance metrics, contextual variables (e.g., climate, orientation), and façade mechanisms.

The literature search was conducted using three major academic databases—Scopus, ScienceDirect, and Google Scholar—chosen for their comprehensive coverage of high-impact studies in architecture, urban design, and environmental science. The keyword selection was based on two sets of queries: first, for thermal, acoustic, and visual comfort; and second, for materials and green façades. The following Boolean operators were used to identify the scope of work in this study:

1. ("outdoor thermal comfort" OR "exterior thermal comfort" OR "external thermal comfort" OR "ambient thermal comfort") AND ("building façade" OR "building envelope" OR "exterior wall")
2. ("outdoor visual comfort" OR "urban visual comfort" OR "visual comfort") AND ("building façade" OR "building envelope" OR "exterior wall")
3. ("urban acoustic" OR "acoustic comfort" OR "outdoor noise") AND ("building façade" OR "building envelope" OR "exterior wall")
4. ("vertical green systems" OR "green walls" OR "living walls" OR "material performance") AND ("building façade" OR "building envelope" OR "exterior wall").

The reviewed articles were selected from works published between 2010 and 2024, with a specific focus on urban outdoor environments and façade strategies that impact at least one domain of OEC. The selection process is illustrated in Figure 1, following a PRISMA-style flow diagram for transparency. The identification of records from the databases included 920 articles. All duplicated articles were removed before screening. In the screening process, titles and abstracts were screened to eliminate unrelated records. Next, all the full papers were screened to ensure the following exclusions were met: (1) focusing on indoor environments; (2) lacking a direct focus on façade materials or green façades; and (3) any non-English-language and non-peer-reviewed publications. Finally, a total of 41 articles meeting these criteria were included in the review. Refer to Figure 1 for the flow diagram on the selection of works.

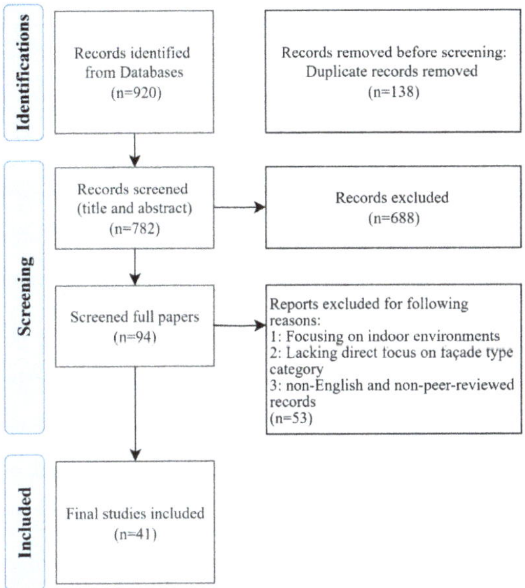

Figure 1. Flow diagram for the selection of works.

After screening and eligibility checks, the selected studies were grouped into three comfort categories: OTC, OAC, and OVC. Within each category, studies were analyzed to assess the role of façade materials and green wall systems in influencing environmental parameters, including temperature, noise levels, and visual glare. Furthermore, the studies were categorized into two levels: building and urban. The building category considers studies near buildings, while the urban category encompasses buildings in urban blocks or neighborhoods. This review was guided by the following research questions (RQs):

- RQ1: What types of materials and vegetation have been investigated in façade design for improving outdoor comfort?
- RQ2: Which performance metrics are most commonly used to assess their impact on OTC, OAC, and OVC?
- RQ3: What measurable outcomes have been reported, and how can these findings inform the design of future climate-responsive façades?

This methodology provides a transparent and structured foundation for synthesizing interdisciplinary research on façade-driven outdoor environmental comfort in urban contexts.

4. Results

Based on 41 articles, a literature review was conducted by grouping the total number of articles into the three OEC domains: OTC, OAC, and OVC. Next, the façade categories, façade materials, and green façades were extracted to create a subgroup from which their types were extracted for each domain. The common OEC metrics used in these studies were then gathered along with their findings. Additionally, studies were then grouped under buildings or urban scale, as another layer of information. All these groupings are synthesized and highlighted in Table 1.

Table 1. Synthesis of review findings.

References	Category	Material/Green Façade Types	Building/Urban Scale	Domain (No. of Studies)	Common Metrics	Summary of Findings
[6,7,10–12,17,38–46]	Material	Brick, Stone, Concrete, RR, DHR, RR, BIPV, HPL, Ceramic, Aluminum, Glass	Building (9)/Urban (14)	OTC (23)	PMV, MRT, PET, UTCI, surface T, WBGT, SET*	OTC is influenced by material properties that have a strong relationship with climate and design parameters. Materials with high reflectivity, light colors, and low thermal mass can help lower surface and air temperatures, thereby reducing the intensity of urban heat islands (UHI). Similarly, green façades can decrease surface temperatures and contribute to lower average physiological equivalent temperature (PET) levels.
[10,13,14,16,18,47–50]	Green Facade	Living Wall System, Green walls with climbing plants				
[8,19,51–56]	Material	Concrete, Brick, Cork Wood, Fiber-cement, Glass, HPL	Building (6)/Urban (7)	OAC (13)	SPL, LAeq, EDT, STI, NR, RT20, D50, Sound Absorption Coefficient	OAC is influenced more by design parameters than by material properties. The geometry of façades and the sound absorption coefficients of materials have a significant impact on urban noise propagation. Green wall systems not only reduce noise but also serve as effective, dual-function interventions that enhance both aesthetics and comfort.
[15,57–60]	Green Facade	Living Wall System, Green walls with climbing plants				
[9,61–64]	Material	N/A	Building (2)/Urban (3)	OVC (5)	DGP, DGI, DG, E_cyl, E_h, E_v	OVC is an under-researched area that closely relates to material properties and design parameters. Certain materials with high specular reflectivity and saturated finishes contribute to glare, visual fatigue, and pupil dilation, which can be mitigated through appropriate design.

The results were based on an analysis structured and conducted at three levels for each of the two categories: materials and green façade, for each of the three comfort factors. The first level involved investigating the "Performance Determinants" through the lens of Material/System Properties, Design Parameters, and Climatic Context, which contributed to OEC. The second involved extracting the popular "Models/Metrics" used for each category. The third level identified the "Material Types" or "Green Façade Types" in the literature. A detailed table of this analysis is found in the Supplementary Materials listing the findings from each article.

4.1. Outdoor Thermal Comfort and Façade Materials

4.1.1. Performance Determinants

The review results highlight the primary and secondary factors that exhibit the relationship between materials in the envelope system and OEC. Primary factors include material properties, while secondary factors include urban design parameters and climatic context.

Material properties as primary factors are direct factors impacting the outdoor environment and include: *solar reflection (SR) & emissivity (e)* [10,11]; *thermal mass* (T_m) [10]; *heat retention* (H_r) [10,38]; *thermal flux* (T_f) [38]; *albedo (A)* [7,11,39]; *solar absorption (SA)* [40]; *retro reflectivity*, where light returns to its source [6,11]; *diffused reflectivity*, where light scatters at different angles [11]; *solar absorbance in Building Integrated Photovoltaic (BIPV)* [41]; and *color & texture* [10,42].

Climatic context and urban design parameters are the secondary factors that indirectly impact the material performance of an envelope system. Climatic contexts include the *Tropical rainforest (Af)* context of Singapore [41]; the *Hot semi-arid (BSh)* context of Iraq [17], Agadir, Morocco [43], Shiraz, Iran [10], Biskara city, Algeria [38] and Errachidia, Morocco [43]; the *Humid subtropical (Cfa)* context of Toyohashi, Japan [11] and Parma, Italy [42]; the *Temperate oceanic (Cfb)* context of London, UK [39] and Zurich, Switzerland [41]; the *Hot summer Mediterranean (Csa)* context of Fez, Morocco [43], and Rome, Italy [6]; the *Humid continental (Dfb)* context of Munich, Germany [40]; and the *Humid Continental (Dwa)* context of Tianjin, China. Urban design parameters include *sky view factor (SVF)* [42]; *aspect ratios* [11,38,43]; *orientation* [38]; *canyon width* [41]; and *urban geometry and building height* [17,41,45].

Urban parameters like aspect ratios, canyon widths, and orientation are essential to UHI, and their connection with façade materials has been extensively studied to understand the OTC levels in these areas. Materials show high heat retention or cooling capacity in specific urban configurations. For instance, while brick can cause a heat retention range varying from 70% to 93% within the same canyons, limestone, in a north–south orientation, can provide a cooling efficiency range from 85% to 100% in deep and shallow canyons [38]. Another study showed that retroreflective (RR) walls demonstrated better performance in higher urban aspect ratios [11]. Other urban parameters include ground surface material, building typology, building height, and canyon width. These parameters were explored in a study finding that ground surface material was the most influential factor, and street width was the most effective factor affecting the OTC levels [41]. Another study showed that façades in high-rise configurations have a higher impact on OTC levels than in low-rise urban settings [17]. Urban geometry also affects solar availability on building façades, impacting the OTC levels of an outdoor urban space. A recent study on 24 urban configurations in London explored the amount of solar availability on façades, highlighting that three urban parameters, namely layout complexity, standard deviation of building height, and directionality, affected the diffuse and direct solar irradiance, thus impacting the OTC levels [45]. A study in Moroccan cities exploring the heat flux from façades with conventional brick walls and historic thick walls showed that medium aspect ratios are best suited to achieve acceptable PMV levels [43].

Physical properties, such as color, texture, emissivity, and solar reflectance, have been crucial in understanding the relationship between building façades and OTC levels. A study examining how physical traits affect PET levels found that materials with high albedo (light-colored cement, travertine, or aluminum) raise the mean radiant temperature and PET index despite having lower surface temperatures [10]. However, a comparative study of two different reflective materials (with high albedo), RR and diffuse highly reflective (DHR) materials with similar solar reflectance showed that, unlike DHR, RR exhibited better performance in reducing thermal environment indicators, namely (Wet-Bulb Globe Temperature (WBGT) & new Standard Effective Temperature SET*); and radiation indicators (sol-air temperature (SAT) & change in operative temperature (COT) [11]. A similar experimental study investigating the albedo values of RR material on façades showed that albedo increased from 38.2% to 42.3%, with the incident angle of the light beam ranging from 8° to 60° angles [6]. It is also found that materials with high thermal mass and low emissivity can result in lower air temperatures [10]. A neighborhood-scale computational study in the humid continental climate of China, exploring the effect of wall albedo change from 0.3 to 0.7, showed a decrease in the wall surface T [7].

4.1.2. Models/Metrics

The review also highlights various performance indicators used to understand the impact of envelope materials on OTC. Two performance indicator categories were used in these studies: (a) OTC metrics such as *PMV* [17,43], *UTCI* [40,41], *SET** & *COT* [11]; *PET* [10,38]; and (b) meteorological variables include *surface temperature (surface T)* [7,10], *air temperature (air T)* [41,42], *direct and diffuse solar radiation* [45]; *mean radiant temperature (MRT)* [10]; *wet-bulb globe temperature (WBGT)*, and *sol-air temperature (SAT)* [11].

4.1.3. Material Types

The review identified two main categories for material types that were studied to evaluate their performance. These are conventional and advanced materials.

Conventional materials: Most research focused on the first category—conventional materials—that included materials such as brick [10,17,38–40,42,43], which was found to be the most widely researched material. This was followed by concrete [10,17,38,40,42]. Other conventional materials included aluminum [10], glass [39], cement [10,40], and different types of stones [10,17,38,42].

Studies have shown that certain materials have a significant impact on OTC levels, particularly in a specific climatic context. A comparative study of twenty commonly used façade materials in sunny and shade conditions in the hot-arid climate of Iran. It showed that façade surface temperature (T_S) is affected by solar reflection, emissivity, and heat capacity properties of materials [10]. The studied materials were brick (cream/red), cement (white), travertine (white/cream/brown/gray), granite (black/gray), ceramic (brown/cream), porcelain (cream/gray), artificial stone (beige), marble (cream/white), aluminum composite (gray), concrete (multicolor), and living wall (continuous/ modular).

Another study in the hot semi-arid climate of Iraq explored five façade materials, namely clay bricks, thermo-stone, glazed panels, granite finish, and concrete blocks, to assess their impact on OTC levels measured through PMV [17]. The findings showed that burnt brick had the worst impact, while thermos-stone and granite walls had the best effect on PMV levels. A study in the hot-dry climate of Biskara, Algeria, investigated four façade materials—hollow brick, concrete hollow block, [adobe] mud brick, and limestone—deriving six material thermal flux categories and highlighting the best material in each case [38]. For instance, limestone was the best material, with a high cooling category (above 60–100%) while brick, concrete, and adobe fell into the extreme heat category.

A large-scale physical model study of a neighborhood in London, comparing the albedo for brick wall, aluminum cladding, and curtain wall, highlighted that the curtain wall, as compared to the aluminum and brickwork façade material, exhibited a lower contribution to the canyon albedo during the day [39], thus implying its poor contribution to acceptable OTC levels in summer. A simulation study of three façade materials, brick, concrete, and plastered insulation with two colors, and their impact on UTCI, conducted in the continental climate of Munich, showed that the concrete material with high solar absorption (80%) had the least UTCI stress, while plastered insulation with low solar absorption (30%) had the highest UTCI stress [40].

While most of these studies showed a strong relationship between materials and OTC, a study on conventional materials stoneware panels (light gray and red colors), high-pressure laminate (HPL) panels (light gray and intense red colors), exposed brick, and concrete in marine climate of Italy showed significantly fewer differences in the simulation results of air T for the case study [42]. The authors highlight that the sky view factor (SVF) of the urban canyon (nearly 1) could have played an important role in these findings.

Advanced materials: Besides conventional materials, there has been an increased interest in advanced materials such as RR [6,11]; DHR [11]; and BIPV [41]. It has been found that reflective materials/coatings in façades help mitigate the UHI effect in urban canyons. There have been studies that explored reflective materials, namely RR material—that returns the light back to the source material—and DHR material—that reflects light at many angles (i.e., scatter light), rather than just one. An experimental study using RR and DHR wall models showed that RR walls showed improved OTC levels (SET* and COT as thermal comfort indices) compared to DHR [11]. However, the Computational Fluid Dynamics (CFD) analysis of the same wall models in this study showed a negligible effect on air temperature.

While high solar reflectance and high solar emittance are ideal material characteristics to reduce the UHI effect, there is a constant rise in the use of BIPV façades to help with onsite energy generation. The inherent solar radiation-absorbing characteristic of BIPV makes it a heat source and thus amplifies the UHI effect when it is used as a façade material. However, their relatively low thermal mass in assembly design helps with quick heating and cooling of the rear-ventilated surfaces. A study on BIPV façades in eighty different urban simulation scenarios showed that the use of BIPV in different climates produces different OTC level ranges. Lower air T ranges result in narrower UTCI ranges, as seen in Singapore, while the opposite is true in Zurich, which has higher ranges [41].

4.2. *Outdoor Thermal Comfort and Green Façades*

4.2.1. Performance Determinants

Green façades impact OTC levels in outdoor environments. Green façade system properties are the primary factors exhibiting a direct relationship between the green façade system and OTC, while urban design parameters and climatic context are the secondary factors, with an indirect relationship. Green façade system properties comprise plant properties, type of plant, leaf area index, leaf angle, plant foliage thickness, percentage of wall coverage, emissivity of substrate, water coefficient of plant substrate, etc., which impact the thermal performance of the green wall [10,14,16,50].

The review highlights urban design parameters as one of two indirect factors impacting the thermal performance of green façades. These include façade orientation [14]; distance from the wall [13,14,49]. The other factor in these studies is climatic context, such as the tropical rainforest (Af) context [50]; the hot semi-arid (BSh) context of Shiraz, Iran [10];

the humid subtropical (Cfa) context of Guangzhou, China [16,50]; the temperate oceanic (Cfb) context [50]; the hot summer Mediterranean continental (Csa) context of Madrid, Spain [13,49,50]; and the hot summer humid continental (Dfa) context of Chicago, USA [14,50].

The findings of this review indicate that green façade systems help reduce PET levels. A comparative study of twenty façade materials highlighted that living walls showed the lowest PET levels and surface temperature (T_s) during the day [10]. The study also highlights that living walls, although expensive, can help mitigate glare problems due to high solar reflectance. Another study on green façades in transitional/shaded areas in the humid subtropical climate of Guangzhou, China, showed that the average PET was reduced by 2.54 °C and 1.43 °C compared to the unshaded area [16].

Moreover, similar results have been observed for microclimatic meteorological variables. For instance, a review study of vertical greening in historic urban context highlights its advantages thanks to its characteristics of reducing fluctuation in microclimate variables, namely surface temperature (T_s), air temperature (T_a), relative humidity (RH), solar irradiation (SR), and particulate matter (PM) [50]. Another study with similar variables in an urban context of a Mediterranean climate showed reduced fluctuations in air temperature (T_a), relative humidity (RH), and solar irradiance (SR) measured at four distances from the green wall [49]. A different climatic context, namely, the humid continental climate of Chicago, showed similar results. The research exploring the effect of a ~20 cm thick layer of Boston ivy (Parthenocissus tricuspidata) on environmental conditions demonstrated that the ivy reduced the surface temperature (T_s) by an average of 0.7 °C and the outdoor air temperature (T_a) by an average of 0.8–2.1 °C depending on the façade orientation [14]. A similar field measurement study of green vs bare walls in the cold semi-arid climate of Madrid, Spain, showed that green walls help with reducing temperatures in outdoor microclimates [13]. The study also demonstrated that reduced air temperature (T_a) in summer was about 2.5–2.9 °C, while in autumn the record was 1.5 °C.

4.2.2. Models/Metrics

The performance indicators used to understand the impact of green façades on the OTC include the OTC metric, namely, *PET* [10,48]; and meteorological variables such as *surface temperature (surface T)* [10,14,50], *globe temperature (globe T)* [48]; *solar reflectance (R_s)* [10]; *air temperature (air T)* [13,49,50]; *relative humidity (RH)* [48–50]; *solar irradiation (SR),* [50]; *wind velocity (W)* [48]; and *mean radiant temperature (MRT)* [48].

4.2.3. Green Façade Types

Green façades are typically categorized into two types based on their design and percentage of coverage—(a) living wall systems with modules, substrate for plants, and a water system (irrigation/drainage) [10,13,49], and (b) green walls with climbing plants on façades [14,16]. These are referred to as 'continuous living walls', 'modular living walls', 'green walls', 'vertical green system, 'ivy façade', etc. [16,18] in the literature. Besides the system design, plant types also play a significant role in their performance with respect to OTC. The list of plant types studied here comprises: Heuchera americana "dale's strain"; Sedum acre "golden carpet; Sedum album "coral carpet"; Thymus communis; Lonicera nitida "maigrun"; Heuchera americana "palace purple"; Carex oshimensis; Delosperma cooperi; Gazania rigens; Thymus vulgaris; and ivy façade with a 20 cm thick layer of Boston ivy (Parthenocissus tricuspidata) [14,49].

4.3. Outdoor Acoustic Comfort and Façade Materials

4.3.1. Performance Determinants

The acoustic performance of building façades plays a crucial role in shaping urban soundscapes by influencing noise levels and enhancing outdoor acoustic comfort. Various studies have examined the effects of different façade materials, geometries, and design strategies on sound absorption, reflection, and diffusion in urban environments. The primary and secondary factors exhibiting the relationship between materials and OAC are material properties and façade design parameters. The material properties in this research are direct factors that impact the acoustic comfort and include sound absorbance (sound reflection) [51,54] and reflectivity [19]. The façade design parameters are the indirect factors impacting the acoustic performance of an envelope system and include *façade geometry details* [51,56], *corner cube retroreflective facades* [55], *louvres* [8], *exposed vs shielded facades* [8], *façade height* [53], *perceiver location* [19], and *balcony design* [8,19,51,54].

Façade design parameters include geometric configurations that significantly impact noise attenuation, with research indicating that incorporating sound-absorbing cladding can reduce the SPL by up to 10 dB on the treated façade and 3 dB on the opposite façade. This highlights the importance of early integration of shape and material optimization in achieving acoustic comfort [51]. Additionally, different materials exhibit varying reflective properties, with clinker brick and mineral plaster producing higher sound reflections compared to HPL hardboard, which has lower reflection values by up to 0.8 dB at a height of 2 m [52].

The role of façade geometry is further emphasized in studies analyzing retroreflection effects, where specific geometries, such as corner cube arrays, resulted in increased retroreflection levels by 5–20 dB compared to flat surfaces, particularly in the 2000–4000 Hz frequency range [55]. These findings underscore the significance of façade geometry in redirecting sound energy and enhancing acoustic comfort in urban environments. Furthermore, a systematic review of building envelope strategies identified that textured façades and sound-absorbing materials could achieve noise-reduction levels ranging from 5.8 dBA in exposed street canyons to 7.3 dBA in shielded areas, with elements such as balconies and louvres contributing to improved sound diffusion and absorption [8]

Experimental research in university courtyards has demonstrated that taller façades with reflective materials, such as concrete and large glass surfaces, resulted in prolonged reverberation times, reducing speech clarity and intelligibility, whereas shorter façades with diffusive materials showed improved acoustic conditions [53]. Similarly, noise mitigation studies in urban street canyons found that materials such as pumice stone and refractory brick provided up to 5.1 dB insertion loss at 5 kHz, while cork wood and metal wool offered lower noise-reduction capabilities, emphasizing the importance of selecting materials based on their frequency-dependent performance [54].

Balcony design is also critical in OAC. For instance, a study shows that balconies behave acoustically like coupled cavity resonators that trap and dissipate part of the incident sound energy [65]. A psychoacoustic study further emphasized that moderate façade absorption (e.g., $\alpha w \approx 0.55$) resulted in optimal outdoor acoustic comfort, while overly reflective or highly absorptive materials caused discomfort. The study highlighted the importance of façade absorption properties in combination with balcony design and sound sources in shaping perceived comfort levels [19]. Recent advancements in computational methods, such as rule-based frameworks and Boundary Element Method (BEM) simulations, have enabled dynamic adaptation of façade geometries to acoustic constraints, revealing that curvilinear surfaces produce better sound scattering compared to planar designs. This approach allows for scalable and efficient acoustic evaluations in urban planning [56].

4.3.2. Models/Metrics

The performance indicators used to understand the impact of façade design on the OAC include sound-pressure level (*SPL*) [51,53], *A-weighted equivalent sound level (LAeq)* [52,54], *reverberation time (RT)* [53], *early decay time (EDT)* [19,53], *speech definition index (D50)* [19,53], *Rapid Speech Transmission Index (RaSTI)* [53], *sound-absorption coefficient* [54], *early reflection SPL* [55], *noise abatement measured through insertion loss* [8], *speech transmission index (STI)* [19], speech clarity index (C50) [19], *music clarity index (C80)* [19], lateral energy fraction (LF80) [19], *perceived comfort* [19], *reflected sound pressure* [56], and *noise reduction* [8].

4.3.3. Material Types

Types of materials explored in OAC include glass [51,53], brick [52,54], plaster [51,52], high-pressure laminate (HPL) [52], concrete [53], pumice stone [54], cork [54], metal wool [54], foam [54], nylon and polyester compound [54], and stainless steel sponge [54]. A comparative analysis of multiple materials conducted in narrow vertical strips (comprising only 8–17% of the wall area) made of porous pumice or refractory brick achieved a 3–5 dB insertion loss across the 1.25–5 kHz band. In contrast, the untreated hard wall provided virtually no mitigation [54]. Another study showed that a rigid laminate (HPL) panel added only 0.8 dB of reflection at 2 m from the wall. In contrast, mineral plaster and clinker brick returned 2.5 dB and 2.8 dB, respectively, confirming that HPL behaves as the most acoustically "soft" of the three claddings [52].

The impact of façade materials on urban acoustics is multifaceted, involving a combination of material properties, geometric configurations, and strategic design interventions. Studies have shown that sound-absorbing materials, geometric adaptations, and integrated design approaches can significantly improve acoustic comfort in urban environments. Future research should continue exploring innovative materials and computational techniques to enhance the acoustic performance of urban façades.

4.4. Outdoor Acoustic Comfort and Green Façades

4.4.1. Performance Determinants

Similar to OTC, green façade system properties serve as the primary factor for OAC performance and comprise plant type, leaf area index, leaf area density, angle of leaf orientation, growing medium, coverage, substrate design (type, thickness, and moisture content), structural design, construction materials, and air layer [15,57,58,60]. For instance, a recent review study has demonstrated the potential to reduce noise levels by up to 40 dB, depending on system design, plant type, and substrate composition [57]. Other studies have highlighted the effectiveness of green façades in reducing noise through variations in vegetation density, substrate properties, and structural design. Green walls with moderate greenery coverage (65–80%) and optimized structural support systems, such as recycled wood-plastic composites, provide significant noise-reduction benefits, particularly in the low-to-mid-frequency range (~1000 Hz) [15] Substrate thickness plays a crucial role in absorption, with thicker substrates achieving noise reductions of 5–10 dB, whereas thinner substrates result in lower reductions (2–3.9 dB) [58]. The presence of specific plant species, such as Bergenia cassifolia and Jacobaea maritima, enhances absorption, with larger leaves contributing more effectively to mid-high frequencies and smaller leaves performing better at high frequencies [59].

Research also indicates that indirect green façades, installed with an air gap between the wall and vegetation, perform better than direct installations, achieving up to 2 dB greater noise reduction [57]. Additionally, studies have highlighted that the moisture content in the substrate can negatively impact performance by filling air pockets and reducing overall absorption efficiency [15]. Field studies have confirmed that modular green wall systems can improve sound absorption across all octave bands, with the best performance observed at mid to high frequencies [60]. Moreover, studies have shown that adding water to the growing medium can negatively impact noise absorption, as it fills the air pockets that contribute to acoustic attenuation [58]. Green walls have demonstrated an absorption coefficient of up to 0.40 in laboratory tests, with the best results seen in the mid-to-high frequency range [59]. In addition, studies conducted in street canyons suggest that full façade coverage is not necessary to achieve effective noise reduction; partial coverage can provide similar results with lower installation costs [57].

On the other hand, the secondary factor design parameters contributing to noise reduction include *canyon width* [57,60], *distance from the source* [58], and *façade width, building heights, water needs, wall types* [60]. For instance, studies have shown that green façades and living wall systems (LWSs) have emerged as practical solutions to mitigate urban noise pollution, a significant challenge that impacts human health and well-being. While green walls offer notable acoustic benefits, their effectiveness varies depending on environmental conditions and system configurations. The cost-effectiveness of these systems for noise mitigation alone has been questioned, emphasizing the need for optimized material selection and design improvements to enhance performance. Furthermore, green façades are most effective in narrow street canyons where they can reduce noise reflections and contribute to overall urban comfort [60].

4.4.2. Models/Metrics

The most common performance metric for acoustic was found to be *noise reduction (NR)* [57]. This was followed by *sound-reflection levels (SRL)* [57], *sound-absorption coefficient (SAC)* [15,57–59], *noise absorption at different frequencies* [58,59], *sound attenuation (SA)* [59], and *sound reduction index* (SRI) [59].

4.4.3. Green Façade Types

Green façades in this section have been categorized into (a) *living wall systems* and (b) *green walls* [57]. While the substrate design and its construction have been identified as contributing factors, the plant type also plays an essential role in its performance. The list of plant types studied in this scope includes Jacobaea maritima [15,59], *Hedera helix* [58], and *Bergenia cassifolia* [57,58], and *Helichrysum thianschanicum* [59]. Overall, green façades are a promising solution for mitigating urban noise, providing measurable reductions across various frequency ranges. To maximize their potential, future research should focus on optimizing substrate properties, vegetation types, and structural configurations to achieve improved acoustic performance and cost efficiency.

4.5. Outdoor Visual Comfort and Façade Materials

4.5.1. Performance Determinants

Envelope materials play a significant role in OVC, and material properties are a direct factor that interacts with the occupants and impacts their OVC. These include *light reflectance* [61,62], *material colors* [63,64,66], and *daylight absorption* [9]. Other design parameters include *age and gender* [61], *sky conditions (rainy and clear skies)* [63], *eye movement* [64], and *sky view factor* [9].

The study of these factors centers on key issues such as glare, visual fatigue, and reduced daylight availability. For instance, studies show that glare often hinders clear vision, causing irritation [9]. Building façades significantly influence pedestrian visual comfort by affecting material reflectance and daylight illuminance [62]. The authors of this survey study explored the reflectance of clear versus dark façades in Fribourg, Switzerland. They found that high-reflectance materials contributed to extreme glare in bright outdoor conditions [62], highly reflective materials on glazed envelope [61], and surfaces with different color attributes [64]. Another survey on highly reflective materials on glazed envelopes showed that reflective façades significantly impact visual discomfort by increasing glare duration and prompting avoidance behavior among pedestrians. The study also showed that males exhibit higher avoidance responses, while females are more sensitive to glare [61]. Thus, field measurements and user surveys, as research methods, have contributed to understanding the role of façade material properties in glare perception and brightness levels. To enhance outdoor visual comfort and reduce glare-related discomfort, incorporating effective shading strategies is recommended.

In addition to reflectivity, façade color significantly influences outdoor visual comfort, especially under varying weather conditions. Clear weather conditions resulted in higher comfort ratings compared to overcast and rainy conditions, highlighting the importance of color selection in optimizing visual comfort [63]. Eye-tracking studies have shown that highly saturated or extremely bright colors increase cognitive load and visual fatigue [64]. Excessively bright or saturated façades can contribute to visual discomfort and increased pupil dilation, underscoring the importance of balanced color application in façade design.

Daylight availability and distribution are also crucial for visual comfort in urban outdoor spaces. Studies have shown that façade materials and urban geometry have a significant influence on daylight distribution. In urban areas, the sky view factor (SVF) also plays a critical role [9]. Higher SVF values enhance daylight penetration, improving visual comfort in less dense areas, while compact layouts with low SVF restrict daylight access, leading to visual discomfort [9]. However, research on the specific impact of building façades on outdoor visual comfort remains limited.

4.5.2. Models/Metrics

Various performance indicators have been identified in the OVC literature. These include *cylindrical illuminance* [62], *specular façade reflectivity (SFR)* [61], *perceived glare* [61,62], *luminous sensation mean vote* [62], *discomfort glare (DG)* [62], *occupant behaviors (wearing sunglasses, moving hands above eyes, rotating or bending head, and blinking* [62], *gaze, and swipe* [64]), *perceived visual comfort (PVC)* [63], *luminance contrast (VC)* [63], *chromaticity contrast (CC)* [63], *color attributes (hue, saturation, brightness)* [64], *daylight glare index* [9], *horizontal illuminance, IES luminance ratios, daylight glare probability (DGP)* [66], and *daylight glare index (DGI)* [9,66]. While the daylight factor is one of the common indicators of visual comfort, limitations persist in its accuracy, as it does not consider the dynamic outdoor variations of daylight environments [63]. While some studies have focused on one metric, others have investigated multiple metrics to assess visual comfort. For instance, the study in Idaho, USA, explored multiple metrics, namely horizontal illuminance, vertical illuminance, IES luminance ratios, DGP, and DGI, and found vertical illuminance to be the best metric [66]. Another study using cylindrical illuminance sensors identified highly reflective materials as major contributors to excessive glare, especially in high-illuminance areas [62]. One of the key outcomes of this study underscores cylindrical illuminance as a tool to analyze the geometry of the "open space"-(the space offering the largest free sky view) and is calculated as

$$E_{cyl} = \pi 2 \times (f_{sky} \times L_{sky} + f_{buildings} \times L_{buildings} + (1 - f_{sky} - f_{buildings}) \times L_{ground})$$

where: L_{sky}, $L_{buildings}$ and L_{ground} are the luminance of the sky, the buildings, and the ground; f_{sky} and $f_{buildings}$ are "form factors" giving the proportion of the sky and the buildings as "seen" from the specific location. Both values fall within the range of 0 to 1.

Overall, the role of building façades in shaping visual comfort in outdoor spaces is multifaceted. Key considerations include reflectivity, color attributes, material properties, and shading strategies. Compared to thermal and acoustic comfort, visual comfort remains underexplored in façade research. More importantly, in this pool of studies, no research was found on green façades, making it a key research gap in this topic. Thus, compared to thermal and acoustic comfort, visual comfort remains underexplored in façade research.

Most studies rely on small-scale experiments or user surveys, with a limited application of physiological tracking tools. Future research should incorporate eye-tracking, EEG-based discomfort monitoring, or VR-based visual simulation to understand how façade color, reflectivity, and geometry influence outdoor visual experience, especially under varying climate and lighting conditions.

Summary of results:

This research offers an in-depth understanding of the latest studies on materials and green wall treatments for façade design, focusing on OTC, OAC, and OVC. Figure 2 is a visual representation of these findings, structured around the three levels of analysis: Performance Determinants, Models/Metrics, and Types. The study's focus is limited to façade treatment at the building level, which impacts the overall outdoor environment at the urban level.

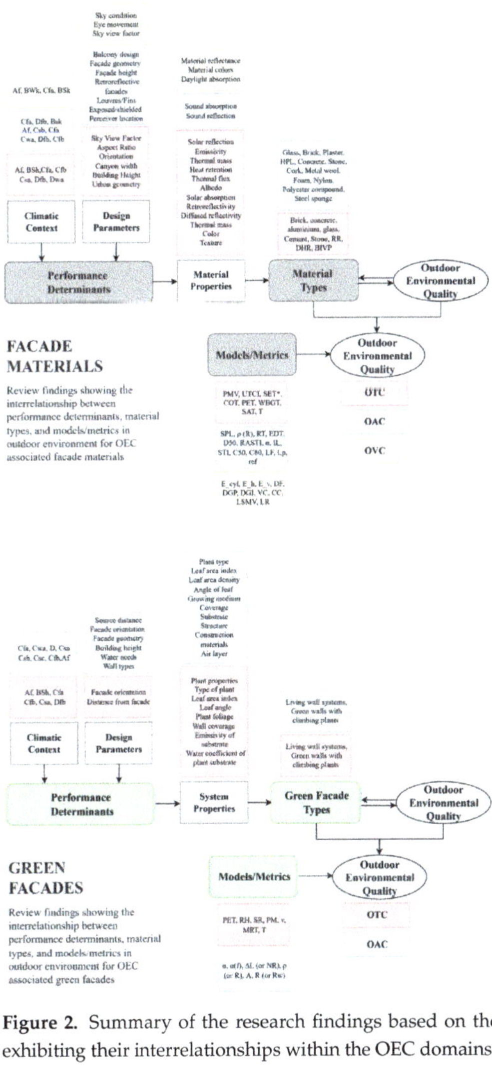

Figure 2. Summary of the research findings based on the three elements of analytical structure exhibiting their interrelationships within the OEC domains for façade materials and green façades.

One of the key findings of this research is the inventory of various factors under each section of the three analytical levels considered for the analysis. This inventory can be used to either develop new materials or assess existing materials in relation to all subsequent categories, rather than just one category. For instance, the performance of brick cannot be evaluated solely by its material properties. It is advisable that the other two categories, namely design parameters and the climatic context, also be considered. Thus, there is an underlying interrelationship between these features of the Performance Determinants. Similarly, extending this further, one must also look into the models/metrics associated with a particular material type.

5. Discussion

This review underscores the critical, yet often underappreciated, role of building façades in shaping OEC. While traditional façade research has primarily focused on indoor environmental quality and energy efficiency, this study shows that façade materials and vegetative treatments greatly impact the thermal, acoustic, and visual quality of nearby outdoor spaces [5]. Through the analysis of 41 peer-reviewed studies, we offer a comprehensive synthesis of the measurable and multifaceted contributions of façades to urban microclimates.

Among the reviewed studies, 24 examined the impact of façade materials or green façades at the urban scale, focusing on urban blocks or neighborhoods. For example, Tabatabaei and Fayaz investigated the effects of 20 façade treatments in a neighborhood in Shiraz [10]. At the building scale, 17 studies focused on immediate surrounding spaces and explored how façade treatments influence OEC. For instance, Wang et al. studied the impact of façade color materials on visual comfort in Shanghai [64]. While building-scale studies have shown façade effects on the immediate microclimate, Prieto and Pastén's "urban acupuncture" approach illustrates that these localized benefits can build up over blocks and neighborhoods. Therefore, we can infer that the interactions between buildings and urban outdoor environmental quality (OEQ) are likely influential in shaping the broader urban–OEQ relationship [67].

To consolidate the review findings and highlight the interaction between façade strategies and outdoor comfort outcomes, Figure 3 presents a conceptual framework. It maps the relationship between two main façade types—material-based and green systems—and the mechanisms through which they influence OTC, OAC, and OVC. These mechanisms include high reflectivity, evapotranspiration, sound absorption, shading, and energy dissipation. By visualizing this interaction, the diagram emphasizes the need for integrative façade design strategies that consider cross-domain impacts.

In the domain of OTC, high-reflectivity materials, light colors, and low thermal mass have demonstrated the capacity to reduce surface and air temperatures, mitigating the intensity of the UHI effect [6,11]. For example, materials such as limestone and retroreflective coatings reduced thermal load more effectively than conventional materials like brick or concrete [10]. Green façades, especially those with continuous vegetative cover, lowered surface temperatures by 2–3 °C [10,14] and contributed to average PET reductions of up to 2.5 °C [16]. However, performance varied across climates and façade orientations, emphasizing the context-dependent nature of these strategies [45,50].

Regarding OAC, studies found that façade geometry and the material's sound-absorption coefficients significantly shape urban noise propagation [52,54]. Retroreflective designs and textured surfaces, such as corner cube arrays or clinker brick, redirected or dispersed sound energy more effectively than flat, reflective façades [52,55]. Green wall systems—especially indirect installations with air gaps—achieved noise reductions of up to 40 dB across specific frequency ranges, making them a powerful dual-function intervention for both aesthetics and comfort [57]. Nonetheless, performance was influenced by factors such as substrate thickness, plant type, moisture content, and installation method [15,59].

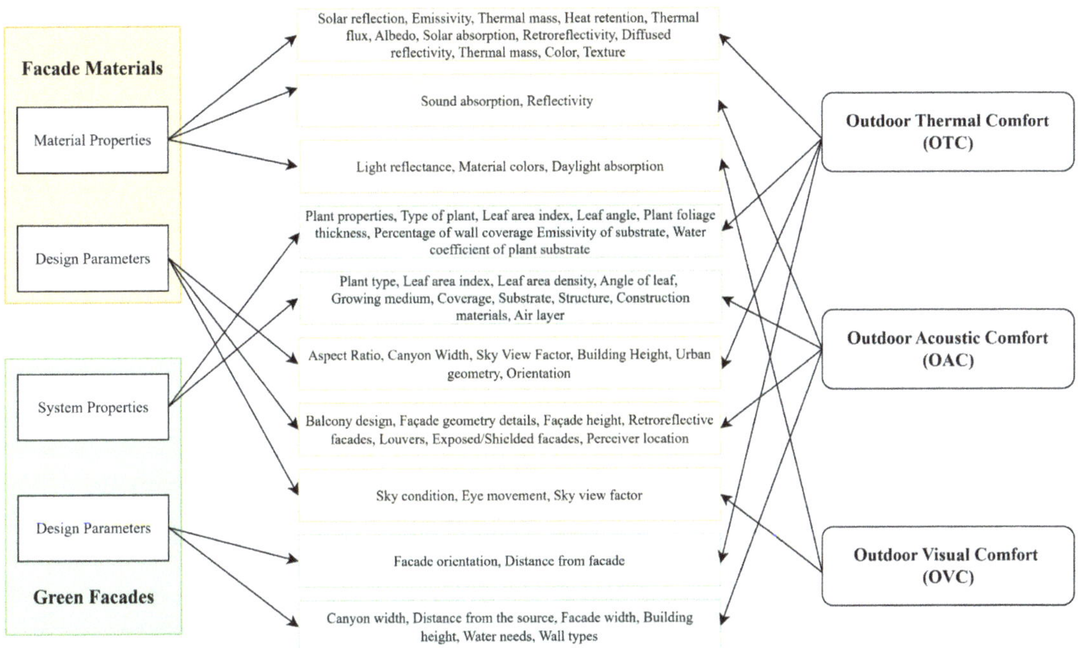

Figure 3. Conceptual diagram linking façade types (materials and green systems) with operative mechanisms (e.g., reflectivity, evapotranspiration, and sound absorption) and their respective impacts on OTC, OAC, and OVC.

In terms of OVC, the literature remains relatively sparse but offers important insights. Materials with high specular reflectivity contributed to glare and behavioral avoidance among pedestrians, particularly under clear sky conditions [61,62]. Studies noted that façades with highly saturated or overly bright finishes increased visual fatigue and pupil dilation [64]. Conversely, strategic shading devices, moderate color palettes, and attention to the sky view factor (SVF) were associated with improved luminance balance and reduced discomfort [9]. While less researched than OTC or OAC, OVC requires increased attention given its implications for user experience and public space quality.

Importantly, our review also identifies a need for cross-domain evaluation, as some façade strategies benefit one domain while potentially compromising another. For instance, while high-albedo surfaces improve OTC by reducing heat gain, they may worsen OVC by increasing glare [68] Similarly, green façades improve both acoustic and thermal comfort but require maintenance and are highly climate-sensitive [55]. These findings necessitate integrative assessments of façade performance, which account for the synergies and trade-offs among thermal, acoustic, and visual metrics.

Quantitative comparisons across studies further illustrate the value of multi-functional façade interventions. Green walls demonstrated temperature reductions of 2–3 °C, noise attenuation of up to 40 dB, and absorption coefficients as high as 0.40 in laboratory settings [59]. Retroreflective façades improved thermal indices like WBGT and UTCI more effectively than diffuse reflective materials [11]. While exact values varied due to geographic, geometric, and climatic differences, these ranges highlight the measurable potential of façade treatments for enhancing OEC [38].

Although these promising findings are encouraging, several limitations are apparent. Our study shows that the geographic distribution of research is heavily concentrated in temperate and arid climates, with fewer assessments in tropical, high-humidity, or high-latitude areas. [34,50]. This limits the generalizability of design recommendations. Moreover, the studies rely on varied comfort indices and modeling tools, making cross-study comparisons difficult [21]. Furthermore, since green façades have shorter lifespans compared to hard materials, we recommend investigating the sensitivity analysis for the durability of various systems [57,59]. However, since the performance and the environmental impact of the same remain higher, it still serves as a sustainable solution for green façades.

To address these gaps and support more robust design practices, we recommend a multi-pronged research agenda. Future work should pursue interdisciplinary methods that simultaneously model thermal, acoustic, and visual comfort outcomes. Long-term field studies are needed to validate simulation results and capture seasonal or maintenance-related effects, particularly for vegetative systems. Additionally, certain studies in the OTC domain have explored advanced computational techniques, such as machine learning and AI-driven generative design for predicting statistical indices [69]. Such approaches can be extended to predict OEC outcomes based on façade properties, providing designers with faster and more holistic performance feedback during early planning phases.

Beyond academic implications, integrating OEC into façade design supports broader sustainability goals. By aligning with policy frameworks such as the United Nations Sustainable Development Goal 11 (Sustainable Cities and Communities) and green building certifications (e.g., LEED, WELL), these strategies can help cities create healthier, more inclusive environments [70,71]. Simulation tools and decision-support systems should advance to incorporate outdoor comfort metrics, in addition to traditional energy and cost factors, to facilitate evidence-based design and ensure regulatory compliance [30,69].

While individual studies often focus on one domain of comfort, façade strategies usually affect multiple parameters simultaneously. Integrated design approaches must weigh the trade-offs to deliver holistic outdoor comfort. The conceptual model presented (Figure 3) supports such assessments. This review advocates for a paradigm shift: from façades as static barriers to façades as dynamic, multi-functional elements that actively shape microclimate, comfort, and public well-being. Through thoughtful material selection, geometric consideration, and vegetative integration, façade design can serve as a powerful lever in building not only energy-efficient but also climatically responsive and socially inclusive urban environments.

6. Conclusions

This review examined 41 peer-reviewed studies published from 2010 to 2024 to assess how façade materials and green systems influence OEC. The study confirms that building façades are not just architectural elements for enclosure or aesthetics but active components that affect outdoor environment quality in thermal, acoustic, and visual aspects.

Out of all the studies, 24 focused on urban-level analysis, while 17 examined individual buildings. This trend highlights a growing focus on the interaction between building façades and OEC. Moreover, the OTC domain was found to be explored more than the OAC and OVC, highlighting further research potential in the latter two domains, particularly OVC. A major finding of this research is the necessity for a broad cross-domain study encompassing OTC, OAC, and OVC, as most existing studies tend to focus on just one domain. Therefore, we developed a conceptual framework to incorporate façade materials into the OEC domains.

Another important finding of this study highlights how various features of the analytical structure (performance determinants, metrics/models, and material/green façade types) are interrelated and could significantly impact the OEC. Specific findings in this area are briefly presented below:

- OTC is impacted by material properties distinctly with close connections to climates and design parameters. High-reflectivity materials, light colors, and low thermal mass could reduce surface and air temperatures, thus mitigating UHI intensity. Similarly, green façades could lower surface temperatures and contribute to average PET reductions around 2.5 °C.
- OAC is impacted by design parameters besides material properties. Façade geometry and the material's sound-absorption coefficients significantly shape urban noise propagation. Green wall systems, while reducing noise up to 40 dB, also serve as powerful, dual-function interventions that enhance both aesthetics and comfort.
- OVC, an under-researched area, shows a close association with material properties and design parameters. Certain materials with high specular reflectivity and saturated finishes can contribute to glare, visual fatigue, and pupil dilation, which can be addressed through proper design.
- One façade treatment (façade material or green façade) could impact all three comfort factors related to OEC, but trade-offs must be evaluated. For instance, high-albedo surfaces could improve OTC by reducing heat gain but worsen OVC by increasing glare.
- Limited work on OEC and façades is available in tropical, high-humidity, or high-latitude regions. Since OEC is climate- and context-specific, additional research is required to cover variations in both design conditions.

The analytical structure of this review paves new research pathways exploring the quantitative interrelationships between the three elements (performance determinants, metrics/models, and material/green façade types). All in all, this review contributes to the existing knowledge on building façade design and its impact on the urban fabric. As metropolitan areas face increasing challenges from climate change, optimizing the performance of building façades is both a design imperative and a policy opportunity. Aligning façade strategies with global frameworks such as the UN SDGs can enhance the livability and resilience of cities. Designers and planners are thus encouraged to consider building envelopes as part of the larger urban ecosystem.

Supplementary Materials: The following supporting information can be downloaded at: https://www.mdpi.com/article/10.3390/buildings15162829/s1, Table S1: Review findings from the list of articles.

Funding: This research received no external funding.

Data Availability Statement: No new data were created or analyzed in this study. Data sharing is not applicable to this article.

Conflicts of Interest: The authors declare no conflicts of interest.

Abbreviations

The following abbreviations are used in this manuscript:

Abbreviation	Metric/Variable
PMV	Predicted Mean Vote
UTCI	Universal Thermal Climate Index
SET*	Standard Effective Temperature (ASHRAE)
COT	Composite Outdoor Thermal Index

Abbreviation	Metric/Variable
PET	Physiological Equivalent Temperature
Tsurf, Ts	Surface temperature
Ta	Air temperature
Sdir	Direct solar radiation
Sdif	Diffuse solar radiation
Tmrt/MRT	Mean radiant temperature
WBGT	Wet-Bulb Globe Temperature
SAT	Sol-air temperature
Tg	Globe temperature
Rs	Solar reflectance
RH	Relative humidity
SR	Solar irradiation
PM	Particulate matter concentration
DHR	Diffuse highly reflective
C50	Speech clarity index (speech)
C80	Music clarity index (music)
D50	Speech definition index
EDT	Early decay time
Eref	Reflected energy levels
Lp,ref	Reflected sound-pressure level
LF80 (or LF)	Lateral energy fraction
RASTI	Rapid Speech Transmission Index
RT	Reverberation time (e.g., RT60, T30)
RR	Retroreflective
ρ (or R)	Sound-reflection coefficient
SPL	Sound-pressure level
STI	Speech Transmission Index
α	Sound-absorption coefficient (fraction of incident energy absorbed)
α(f)	Frequency-dependent sound-absorption coefficient
NR)	Noise-reduction value
ρ (or R)	Sound-reflection coefficient/reflection level
A	Sound attenuation (dB loss through a medium or layer)
R (or Rw)	Sound-reduction index (frequency-band or weighted overall value)
Ecyl	Cylindrical illuminance
Eh	Horizontal illuminance
Ev	Vertical illuminance
DF	Daylight factor
DGP	Daylight glare probability
DGI	Daylight glare index
VC	Luminance contrast
CC	Chromaticity contrast
LSMV	Luminous sensation mean vote
LR	IES luminance ratio

References

1. Masson-Delmotte, V.; Zhai, P.; Pirani, A.; Connors, S.L.; Péan, C.; Berger, S.; Caud, N.; Chen, Y.; Goldfarb, L.; Gomis, M.I.; et al. (Eds.) Climate Change 2021: The Physical Science Basis. In *Contribution of Working Group I to the Sixth Assessment Report of the Intergovernmental Panel on Climate Change*; Technical Summary; Cambridge University Press: Cambridge, UK; New York, NY, USA, 2021. Available online: https://www.ipcc.ch/report/ar6/wg1/downloads/report/IPCC_AR6_WGI_FullReport.pdf (accessed on 26 January 2025).
2. United Nations. Population. United Nations. Available online: https://www.un.org/en/global-issues/population (accessed on 26 January 2025).

3. Zeng, F.; Liang, S.; Zhang, J.; Chen, Y.; Feng, S.; Mo, Q.; Zhou, T.; Lai, Y.; Liu, T.; Wang, S. The effects of acoustic-light-thermal environment quality parameters on pedestrians' overall comforts in residential districts. *Sci. Rep.* **2024**, *14*, 19625. [CrossRef]
4. Whyte, W.H. *The Social Life of Small Urban Spaces*; Conservation Foundation: Washington, DC, USA, 1980.
5. Gupta, V.; Deb, C. Envelope design for low-energy buildings in the tropics: A review-ScienceDirect. *Renew. Sustain. Energy Rev.* **2023**, *186*, 113650. [CrossRef]
6. Battista, G.; de Lieto Vollaro, E.; Grignaffini, S.; Ocłoń, P.; Vallati, A. Experimental investigation about the adoption of high reflectance materials on the envelope cladding on a scaled street canyon. *Energy* **2021**, *230*, 120801. [CrossRef]
7. Liu, Y.; Chu, C.; Zhang, R.; Chen, S.; Xu, C.; Zhao, D.; Meng, C.; Ju, M.; Cao, Z. Impacts of high-albedo urban surfaces on outdoor thermal environment across morphological contexts: A case of Tianjin, China. *Sustain. Cities Soc.* **2024**, *100*, 105038. [CrossRef]
8. Yang, W.; Jeon, J.Y. Design strategies and elements of building envelope for urban acoustic environment. *Build. Environ.* **2020**, *182*, 107121. [CrossRef]
9. Nasrollahi, N.; Shokri, E. Daylight illuminance in urban environments for visual comfort and energy performance. *Renew. Sustain. Energy Rev.* **2016**, *66*, 861–874. [CrossRef]
10. Tabatabaei, S.S.; Fayaz, R. The effect of facade materials and coatings on urban heat island mitigation and outdoor thermal comfort in hot semi-arid climate. *Build. Environ.* **2023**, *243*, 110701. [CrossRef]
11. Yuan, J.; Shimazaki, Y.; Zhang, R.; Masuko, S.; Cao, S.-J. Can retro-reflective materials replace diffuse highly reflective materials for urban buildings' wall to improve outdoor thermal comfort? *Heliyon* **2023**, *9*, e14872. [CrossRef] [PubMed]
12. Yuan, J.; Masuko, S.; Shimazaki, Y.; Yamanaka, T.; Kobayashi, T. Evaluation of outdoor thermal comfort under different building external-wall-surface with different reflective directional properties using CFD analysis and model experiment. *Build. Environ.* **2022**, *207*, 108478. [CrossRef]
13. de Jesus, M.P.; Lourenço, J.M.; Arce, R.M.; Macias, M. Green façades and in situ measurements of outdoor building thermal behaviour. *Build. Environ.* **2017**, *119*, 11–19. [CrossRef]
14. Susorova, I.; Azimi, P.; Stephens, B. The effects of climbing vegetation on the local microclimate, thermal performance, and air infiltration of four building facade orientations. *Build. Environ.* **2014**, *76*, 113–124. [CrossRef]
15. Chang, L.-T.; Chang, F.-C. Study of Living Wall Systems' (LWSs) Support system for improving LWSs Life cycle performance and noise reduction potential. *Build. Environ.* **2022**, *216*, 109007. [CrossRef]
16. Lin, H.; Xiao, Y.; Musso, F.; Lu, Y. Green Façade Effects on Thermal Environment in Transitional Space: Field Measurement Studies and Computational Fluid Dynamics Simulations. *Sustainability* **2019**, *11*, 5691. [CrossRef]
17. Abrahem, S.A.; Hassan, S.A.; Khafoxmees, W.A. Impact of Façade Material of Mass Housing on Outdoor Thermal Comfort in Hot-arid Climate. *IOP Conf. Ser. Mater. Sci. Eng.* **2020**, *881*, 12006. [CrossRef]
18. Perini, K.; Ottelé, M.; Haas, E.M.; Raiteri, R. Vertical greening systems, a process tree for green façades and living walls. *Urban Ecosyst.* **2013**, *16*, 265–277. [CrossRef]
19. Taghipour, A.; Athari, S.; Gisladottir, A.; Sievers, T.; Eggenschwiler, K. Room Acoustical Parameters as Predictors of Acoustic Comfort in Outdoor Spaces of Housing Complexes. *Front. Psychol.* **2020**, *11*, 344. [CrossRef]
20. *ANSI/ASHRAE Standard 55-2023*; Thermal Environment Conditions for Human Occupancy. ASHRAE: Atlanta, GA, USA, 2023. Available online: https://www.ashrae.org/technical-resources/bookstore/standard-55-thermal-environmental-conditions-for-human-occupancy (accessed on 13 July 2024).
21. de Freitas, C.R.; Grigorieva, E.A. A comprehensive catalogue and classification of human thermal climate indices. *Int. J. Biometeorol.* **2015**, *59*, 109–120. [CrossRef] [PubMed]
22. de Freitas, C.R.; Grigorieva, E.A. A comparison and appraisal of a comprehensive range of human thermal climate indices. *Int. J. Biometeorol.* **2017**, *61*, 487–512. [CrossRef]
23. Höppe, P. The physiological equivalent temperature–A universal index for the biometeorological assessment of the thermal environment. *Int. J. Biometeorol.* **1999**, *43*, 71–75. [CrossRef]
24. Fiala, D.; Havenith, G.; Bröde, P.; Kampmann, B.; Jendritzky, G. UTCI-Fiala multi-node model of human heat transfer and temperature regulation. *Int. J. Biometeorol.* **2012**, *56*, 429–441. [CrossRef]
25. Havenith, G.; Fiala, D.; Błazejczyk, K.; Richards, M.; Bröde, P.; Holmér, I.; Rintamaki, H.; Benshabat, Y.; Jendritzky, G. The UTCI-clothing model. *Int. J. Biometeorol.* **2012**, *56*, 461–470. [CrossRef]
26. Bröde, P.; Fiala, D.; Błażejczyk, K.; Holmér, I.; Jendritzky, G.; Kampmann, B.; Tinz, B.; Havenith, G. Deriving the operational procedure for the Universal Thermal Climate Index (UTCI). *Int. J. Biometeorol.* **2012**, *56*, 481–494. [CrossRef]
27. Jendritzky, G.; de Dear, R.; Havenith, G. UTCI—Why another thermal index? *Int. J. Biometeorol.* **2012**, *56*, 421–428. [CrossRef] [PubMed]
28. Fanger, P.O. *Thermal Comfort: Analysis and Applications in Environmental Engineering*; McGraw Hill: New York, NY USA, 1972.
29. Gagge, A.P.; Fobelets, A.P.; Berglund, L. A standard predictive index of human response to the thermal environment. *ASHRAE* **1986**, *92*, 23.

30. Azar, E.; O'Brien, W.; Carlucci, S.; Hong, T.; Sonta, A.; Kim, J.; Andargie, M.S.; Abuimara, T.; El Asmar, M.; Jain, R.K.; et al. Simulation-aided occupant-centric building design: A critical review of tools, methods, and applications. *Energy Build.* **2020**, *224*, 110292. [CrossRef]
31. Verma, A.; Gupta, V.; Nihar, K.; Jana, A.; Jain, R.K.; Deb, C. Tropical climates and the interplay between IEQ and energy consumption in buildings: A review. *Build. Environ.* **2023**, *242*, 110551. [CrossRef]
32. Zarei, F.; Nik-Bakht, M.; Lee, J.; Zarei, F. Urban-Scale Acoustic Comfort Map: Fusion of Social Inputs, Noise Levels, and Citizen Comfort in Open GIS. *Processes* **2024**, *12*, 2864. [CrossRef]
33. Chen, K.; Kang, J.; Ma, H. Evaluation of healthy indoor acoustic environments in residential buildings by the occupants: A mixed-method approach. *Build. Environ.* **2023**, *246*, 110950. [CrossRef]
34. Balderrama, A.; Kang, J.; Prieto, A.; Luna-Navarro, A.; Arztmann, D.; Knaack, U. Effects of Façades on Urban Acoustic Environment and Soundscape: A Systematic Review. *Sustainability* **2022**, *14*, 9670. [CrossRef]
35. *BS EN 12665:2024*; Light and Lighting. Basic Terms and Criteria for Specifying Lighting Requirements. BSI: Brussels, Belgium, 2024. Available online: https://www.en-standard.eu/bs-en-12665-2024-light-and-lighting-basic-terms-and-criteria-for-specifying-lighting-requirements/ (accessed on 27 May 2025).
36. Ma, G.; Pan, X. Research on a Visual Comfort Model Based on Individual Preference in China through Machine Learning Algorithm. *Sustainability* **2021**, *13*, 7602. [CrossRef]
37. Bian, Y.; Luo, T. Investigation of visual comfort metrics from subjective responses in China: A study in offices with daylight. *Build. Environ.* **2017**, *123*, 661–671. [CrossRef]
38. Ben Ratmia, F.Z.; Ahriz, A.; Santi, G.; Bouzaher, S.; Mahar, W.A.; Ben Ratmia, M.A.E.; Matallah, M.E. Street Design Strategies Based on Spatial Configurations and Building External Envelopes in Relation to Outdoor Thermal Comfort in Arid Climates. *Sustainability* **2024**, *16*, 221. [CrossRef]
39. Kotopouleas, A.; Giridharan, R.; Nikolopoulou, M.; Watkins, R.; Yeninarcilar, M. Experimental investigation of the impact of urban fabric on canyon albedo using a 1:10 scaled physical model. *Sol. Energy* **2021**, *230*, 449–461. [CrossRef]
40. Chokhachian, A.; Santucci, D.; Auer, T. A Human-Centered Approach to Enhance Urban Resilience, Implications and Application to Improve Outdoor Comfort in Dense Urban Spaces. *Buildings* **2017**, *7*, 113. [CrossRef]
41. Duran, A.; Waibel, C.; Schlueter, A. A parametric approach to evaluate the impact of BIPV facades on outdoor thermal comfort in different urban contexts. In Proceedings of the Building Simulation 2023: 18th Conference of IBPSA, Shanghai, China, 4–6 September 2023; pp. 1665–1672. [CrossRef]
42. Fabbri, K.; Gaspari, J.; Bartoletti, S.; Antonini, E. Effect of facade reflectance on outdoor microclimate: An Italian case study. *Sustain. Cities Soc.* **2020**, *54*, 101984. [CrossRef]
43. Jihad, A.S.; Tahiri, M. Modeling the urban geometry influence on outdoor thermal comfort in the case of Moroccan microclimate. *Urban Clim.* **2016**, *16*, 25–42. [CrossRef]
44. Fox, J.; Osmond, P.; Peters, A. The Effect of Building Facades on Outdoor Microclimate—Reflectance Recovery from Terrestrial Multispectral Images Using a Robust Empirical Line Method. *Climate* **2018**, *6*, 56. [CrossRef]
45. Chatzipoulka, C.; Compagnon, R.; Nikolopoulou, M. Urban geometry and solar availability on façades and ground of real urban forms: Using London as a case study. *Sol. Energy* **2016**, *138*, 53–66. [CrossRef]
46. Larikova, I.; Fleckenstein, J.; Chokhachian, A.; Auer, T.; Weisser, W.; Dörfler, K. Additively Manufactured Urban Multispecies Façades for Building Renovation. *J. Facade Des. Eng.* **2022**, *10*, 105–126. [CrossRef]
47. Susorova, I.; Angulo, M.; Bahrami, P.; Stephens, B. A model of vegetated exterior facades for evaluation of wall thermal performance. *Build. Environ.* **2013**, *67*, 1–13. [CrossRef]
48. Lin, H.; Chokhachian, A.; Musso, F.; Xiao, Y. Effects of the Green Façade on Thermal Comfort in the Transitional Spaces: Field Measurements in Munich, Germany. In Proceedings of the 5th International Conference on Countermeasures to Urban Heat Islands, Hyderabad, India, 2–4 December 2019.
49. Cosola, V.O.-D.; Olivieri, F.; Olivieri, L.; Ruiz-García, L. Assessment of the impact of green walls on urban thermal comfort in a Mediterranean climate. *Energy Build.* **2023**, *296*, 113375. [CrossRef]
50. De Groeve, M.; Kale, E.; Godts, S.; Orr, S.; De Kock, T. Impact of vertical greening on urban microclimate and historic building materials: A meta-analysis. *Build. Environ.* **2024**, *253*, 111365. [CrossRef]
51. Badino, E.; Manca, R.; Shtrepi, L.; Calleri, C.; Astolfi, A. Effect of façade shape and acoustic cladding on reduction of leisure noise levels in a street canyon. *Build. Environ.* **2019**, *157*, 242–256. [CrossRef]
52. Zagubień, A.; Wolniewicz, K. Measurements and Analysis of Sound Reflections from Selected Building Façades. *Appl. Sci.* **2024**, *14*, 11627. [CrossRef]
53. Hamouta, S.; Zemmouri, N.; Ahriz, A. Facade Design and the Outdoor Acoustic Environment: A Case Study at Batna 1 University. *Buildings* **2024**, *14*, 3339. [CrossRef]
54. Giraldo, J.R.; Colorado, H.A. Evaluation of Noise Mitigation by Different Materials and Balcony Configurations in Urban Street Canyon Facades: Casework in Aburrá Valley, Colombia. *J. Arch. Eng.* **2024**, *30*, 4024012. [CrossRef]

55. Cabrera, D.; Yadav, M.; Holmes, J.; Fong, O.; Caldwell, H. Incidental acoustic retroreflection from building façades: Three instances in Berkeley, Sydney and Hong Kong. *Build. Environ.* **2020**, *172*, 106733. [CrossRef]
56. Tokaç, I.; Heimes, A.; Vorländer, M.; Brell-Çokcan, S. A rule-based framework for capturing geometric characteristics in design: A study of façade analysis for acoustic behaviour in urban space. *Int. J. Arch. Comput.* **2024**, *23*, 405–425. [CrossRef]
57. Radić, M.; Dodig, M.B.; Auer, T. Green Facades and Living Walls—A Review Establishing the Classification of Construction Types and Mapping the Benefits. *Sustainability* **2019**, *11*, 4579. [CrossRef]
58. Cosola, V.O.-D.; Olivieri, F.; Ruiz-García, L. A systematic review of the impact of green walls on urban comfort: Temperature reduction and noise attenuation. *Renew. Sustain. Energy Rev.* **2022**, *162*, 112463. [CrossRef]
59. Ascione, F.; De Masi, R.F.; Mastellone, M.; Ruggiero, S.; Vanoli, G.P. Green Walls, a Critical Review: Knowledge Gaps, Design Parameters, Thermal Performances and Multi-Criteria Design Approaches. *Energies* **2020**, *13*, 2296. [CrossRef]
60. Fonseca, F.; Paschoalino, M.; Silva, L. Health and Well-Being Benefits of Outdoor and Indoor Vertical Greening Systems: A Review. *Sustainability* **2023**, *15*, 4107. [CrossRef]
61. Ishak, N.M.; Jenatabadi, H.S.; Mustafa, S.N.A.M.; Abdullah, J. The Impact of Building Façade Reflectivity on Pedestrian Visual Comfort with the Application of Bayesian Structural Equation Modeling. *Eng. J.* **2021**, *25*, 211–219. [CrossRef]
62. Compagnon, R.; Goyette-Pernot, J. Visual Comfort in Urban Spaces. In *Designing Open Spaces in the Urban Environment: A Bioclimatic Approach*; Center for Renewable Energy Sources: Attiki, Greece, 2004; pp. 27–31.
63. Li, Y.; Xu, B.; Liu, Y. A Study on the Visual Comfort of Urban Building Colors under Overcast and Rainy Weather. *Buildings* **2024**, *14*, 1552. [CrossRef]
64. Wang, Z.; Shen, M.; Huang, Y. Exploring the Impact of Facade Color Elements on Visual Comfort in Old Residential Buildings in Shanghai: Insights from Eye-Tracking Technology. *Buildings* **2024**, *14*, 1758. [CrossRef]
65. Bin Tan, L.; Ang, L.Y.L. Performance Evaluation of Balcony Designs for Mitigating Ground Level Noise. *Acoustics* **2024**, *6*, 272–297. [CrossRef]
66. van den Wymelenberg, K.; Inanici, M. A Critical Investigation of Common Lighting Design Metrics for Predicting Human Visual Comfort in Offices with Daylight. *Leukos* **2014**, *10*, 145–164. [CrossRef]
67. Prieto, A.; Pastén, M. What Is Your Building Doing for the City? Systematic Literature Review on the Potential of Façade Design for the Mitigation of Urban Environmental Problems. *Sustainability* **2024**, *16*, 7855. [CrossRef]
68. Yang, J.; Wang, Z.-H.; Kaloush, K.E. Environmental impacts of reflective materials: Is high albedo a 'silver bullet' for mitigating urban heat island? *Renew. Sustain. Energy Rev.* **2015**, *47*, 830–843. [CrossRef]
69. Guo, R.; Yang, B.; Guo, Y.; Li, H.; Li, Z.; Zhou, B.; Hong, B.; Wang, F. Machine learning-based prediction of outdoor thermal comfort: Combining Bayesian optimization and the SHAP model. *Build. Environ.* **2024**, *254*, 111301. [CrossRef]
70. Worden, K.; Hazer, M.; Pyke, C.; Trowbridge, M. Using LEED green rating systems to promote population health. *Build. Environ.* **2020**, *172*, 106550. [CrossRef]
71. Singh, A.; Chandra, T.; Mathur, S.; Mathur, J. Spatio-temporal assessment of outdoor thermal comfort dynamics in residential neighbourhoods using the TOPSIS method: A novel approach for arid climates. *Sustain. Cities Soc.* **2025**, *126*, 106379. [CrossRef]

MDPI AG
Grosspeteranlage 5
4052 Basel
Switzerland
Tel.: +41 61 683 77 34

Buildings Editorial Office
E-mail: buildings@mdpi.com
www.mdpi.com/journal/buildings

www.ingramcontent.com/pod-product-compliance
Lightning Source LLC
Chambersburg PA
CBHW061547210926
51850CB00009B/2023